Delmar Learning's Test Preparation Series

Automobile Test

Parts Specialist (Test P2)

3rd Edition

THOMSON
DELMAR LEARNING

Australia Canada Mexico Singapore Spain United Kingdom United States

Delmar Learning's ASE Test Preparation Series
Automobile Test for Parts Specialist (Test P2), 3e

Vice President, Technology and Trades SBU:
Alar Elken

Executive Director, Professional Business Unit:
Greg Clayton

Product Development Manager:
Timothy Waters

Developmental Editor:
Christopher Shortt

Channel Manager:
Beth A. Lutz

Marketing Specialist:
Brian McGrath

Production Director:
Mary Ellen Black

Production Manager:
Larry Main

Production Editor:
Elizabeth Hough

Editorial Assistant:
Kristen Shenfield

Cover Images Courtesy of:
DaimlerChrysler Corporation

Cover Designer:
Michael Egan

COPYRIGHT 2004 by Delmar Learning, a division of Thomson Learning, Inc. Thomson Learning™ is a trademark used herein under license.

Printed in Canada
1 2 3 4 5 XX 05 04 03

For more information contact
Delmar Learning
Executive Woods
5 Maxwell Drive, PO Box 8007,
Clifton Park, NY 12065-8007
Or find us on the World Wide Web at :
www.delmarlearning.com, or www.trainingbay.com

ALL RIGHTS RESERVED. No part of this work covered by the copyright hereon may be reproduced in any form or by any means—graphic, electronic, or mechanical, including photocopying, recording, taping, Web distribution, or information storage and retrieval systems—without the written permission of the publisher.

For permission to use material from the text or product, contact us by
Tel.   (800) 730-2214
Fax   (800) 730-2215
www.thomsonrights.com

ISBN: 1-4018-2048-4

NOTICE TO THE READER

Publisher does not warrant or guarantee any of the products described herein or perform any independent analysis in connection with any of the product information contained herein. Publisher does not assume, and expressly disclaims, any obligation to obtain and include information other than that provided to it by the manufacturer.

The reader is expressly warned to consider and adopt all safety precautions that might be indicated by the activities herein and to avoid all potential hazards. By following the instructions contained herein, the reader willingly assumes all risks in connection with such instructions.

The publisher makes no representation or warranties of any kind, including but not limited to, the warranties of fitness for particular purpose or merchantability, nor are any such representations implied with respect to the material set forth herein, and the publisher takes no responsibility with respect to such material. The publisher shall not be liable for any special, consequential, or exemplary damages resulting, in whole or part, from the readers' use of, or reliance upon, this material.

Contents

Preface v

Section 1 The History of ASE

History 1
   ASE 1

Section 2 Take and Pass Every ASE Test

ASE Testing 3
   Who Writes the Questions? 3
   Objective Tests 4
   Preparing for the Exam 5
   During the Test. 5
   Your Test Results! 6

Section 3 Types of Questions on an ASE Exam

   Multiple-Choice Questions 7
   EXCEPT Questions 8
   Technician A, Technician B Questions 8
   Most-Likely Questions 9
   LEAST-Likely Questions 10
   Summary 10
Testing Time Length 11

Section 4 Overview of the Task List

Automobile Parts Specialist (Test P2) 13
   Task List and Overview 14
      A. General Operations (10 Questions) 14
      B. Customer Relations and Sales Skills (11 Questions) 17

  C. Vehicle Systems Knowledge (40 Questions)................20
    1. Engine Mechanical Parts (3 Questions)..............20
    2. Cooling Systems (2 Questions) 24
    3. Fuel Systems (3 Questions) 27
    4. Ignition Systems (3 Questions) 31
    5. Exhaust Systems (2 Questions) 35
    6. Emissions Control Systems (4 Questions) 37
    7. Manual Transmission/Transaxle (2 Questions) 38
    8. Automatic Transmission/Transaxle (2 Questions)......41
    9. Drivetrain Components (2 Questions) 43
    10. Brakes (3 Questions) 47
    11. Suspension and Steering (3 Questions)..............51
    12. Heating and Air Conditioning (3 Questions) 56
    13. Electrical Systems (2 Questions) 57
    14. Battery, Charging, and Starting Systems
     (3 Questions)....................................58
    15. Miscellaneous (3 Questions) 59
  D. Vehicle Identification (3 Questions)......................64
  E. Cataloging Skills (7 Questions)..........................65
  F. Inventory Management (2 Questions)....................66
  G. Merchandising (2 Questions)...........................67

Section 5 Sample Test for Practice

Sample Test..71

Section 6 Additional Test Questions for Practice

Additional Test Questions......................................91

Section 7 Appendices

Answers to the Test Questions for the Sample Test Section 5 115
Explanations to the Answers for the Sample Test Section 5 116
Answers to the Test Questions for the Additional Test Questions
 Section 6..127
Explanations to the Answers for the Additional Test Questions
 Section 6..128

Glossary 143

Preface

Delmar Learning is very pleased that you have chosen our ASE Test Preparation Series to prepare yourself for the automotive ASE Examination. These guides are available for all of the automotive areas including A1–A8, the L1 Advanced Diagnostic Certification, the P2 Parts Specialist, the C1 Service Consultant and the X1 Undercar Specialist. These guides are designed to introduce you to the Task List for the test you are preparing to take, give you an understanding of what you are expected to be able to do in each task, and take you through sample test questions formatted in the same way the ASE tests are structured. If you have a basic working knowledge of the discipline you are testing for, you will find the Delmar Learning's ASE Test Preparation Series to be an excellent way to understand the "must know" items to pass the test. These books are not textbooks. Their objective is to prepare the technician who has the requisite experience and schooling to challenge ASE testing. It cannot replace the hands-on experience or the theoretical knowledge required by ASE to master vehicle repair technology. If you are unable to understand more than a few of the questions and their explanations in this book, it could be that you require either more shop-floor experience or further study. Some textbooks that can assist you with further study are listed on the rear cover of this book.

Each book begins with an item by item overview of the ASE Task List with explanations of the minimum knowledge you must possess to answer questions related to the task. Following that there are 2 sets of sample questions followed by an answer key to each test and an explanation of the answers to each question. A few of the questions are not strictly ASE format but were included because they help teach a critical concept that will appear on the test. We suggest that you read the complete Task List Overview before taking the first sample test. After taking the first test, score yourself and read the explanation to any questions that you were not sure about, including the questions you answered correctly. Each test question has a reference back to the related task or tasks that it covers. This will help you to go back and read over any area of the task list that you are having trouble with. Once you are satisfied that you have all of your questions answered from the first sample test, take the additional tests and check them. If you pass these tests, you will do well on the ASE test.

Our Commitment to Excellence

The 3rd edition of Delmar Learning's ASE Test Preparation Series has been through a major revision with extensive updates to the ASE's task lists, test questions, and accuracy. Delmar Learning has sought out the best technicians in the country to help with the updating and revision of each of the books in the series.

About the Series Editor

To promote consistency throughout the series, a series advisor took on the task of reading, editing, and helping each of our experts give each book the highest level of accuracy possible. Donny Seyfer has served in the role of Series Advisor for the 3rd edition of the ASE Test Preparation Series. Donny brings to the series several years of experience in writing ASE style questions. Donny is an ASE Master, L1 and C1 certified technician, and service consultant. In 2000 and 2001 Donny received a Regional Technician of the Year award. Donny served as a technical member on several automotive boards. Donny is also the host of an auto care radio show and manages his family repair business in Colorado. Additionally, he revised two of the books in this series and wrote the C1 Service Consultant book.

Thanks for choosing Delmar Learning's ASE Test Preparation Series. All of the writers, editors, Delmar Staff, and myself have worked very hard to make this series second to none. I know you are going to find this book accurate and easy to work with. It is our objective to constantly improve our product at Delmar by responding to feedback. If you have any questions concerning the books in this series, you can email me at: autoexpert@trainingbay.com.

Donny Seyfer
Series Advisor

The History of ASE

History

Originally known as The National Institute for Automotive Service Excellence (NIASE), today's ASE was founded in 1972 as a nonprofit, independent entity dedicated to improving the quality of automotive service and repair through the voluntary testing and certification of automotive technicians. Until that time, consumers had no way of distinguishing between competent and incompetent automotive mechanics. In the mid-1960s and early 1970s, efforts were made by several automotive industry affiliated associations to respond to this need. Though the associations were nonprofit, many regarded certification test fees merely as a means of raising additional operating capital. Also, some associations, having a vested interest, produced test scores heavily weighted in the favor of its members.

From these efforts a new independent, nonprofit association, the National Institute for Automotive Service Excellence (NIASE), was established. In early NIASE tests, Mechanic A, Mechanic B type questions were used. Over the years the trend has not changed, but in mid-1984 the term was changed to Technician A, Technician B to better emphasize sophistication of the skills needed to perform successfully in the modern motor vehicle industry. In certain tests the term used is Estimator A/B, Painter A/B, or Parts Specialist A/B. At about that same time, the logo was changed from "The Gear" to "The Blue Seal," and the organization adopted the acronym ASE for Automotive Service Excellence.

ASE

ASE's mission is to improve the quality of vehicle repair and service in the United States through the testing and certification of automotive repair technicians. Prospective candidates register for and take one or more of ASE's many exams.

Upon passing at least one exam and providing proof of two years of related work experience, the technician becomes ASE certified. A technician who passes a series of exams earns ASE Master Technician status. An automobile technician, for example, must pass eight exams for this recognition.

The exams, conducted twice a year at over seven hundred locations around the country, are administered by American College Testing (ACT). They stress real-world diagnostic and repair problems. Though a good knowledge of theory is helpful to the technician in answering many of the questions, there are no questions specifically on theory. Certification is valid for five years. To retain certification, the technician must be retested to renew his or her certificate.

The automotive consumer benefits because ASE certification is a valuable yardstick by which to measure the knowledge and skills of individual technicians, as well as their commitment to their chosen profession. It is also a tribute to the repair facility employing ASE certified technicians. ASE certified technicians are permitted to wear blue and white ASE shoulder insignia, referred to as the "Blue Seal of Excellence," and carry credentials listing their areas of expertise. Often employers display their technicians' credentials in the customer waiting area. Customers look for facilities that display ASE's Blue Seal of Excellence logo on outdoor signs, in the customer waiting area, in the telephone book (Yellow Pages), and in newspaper advertisements.

To become ASE certified, contact:

National Institute for Automotive Service Excellence
101 Blue Seal Drive S.E.
Suite 101
Leesburg, VA 20175
Telephone 703-669-6600
FAX 703-669-6123
www.ase.com

Take and Pass Every ASE Test

ASE Testing

Participating in an Automotive Service Excellence (ASE) voluntary certification program gives you a chance to show your customers that you have the "know-how" needed to work on today's modern vehicles. The ASE certification tests allow you to compare your skills and knowledge to the automotive service industry's standards for each specialty area.

If you are the "average" automotive technician taking this test, you are in your mid-thirties and have not attended school for about fifteen years. That means you probably have not taken a test in many years. Some of you, on the other hand, have attended college or taken postsecondary education courses and may be more familiar with taking tests and with test-taking strategies. There is, however, a difference in the ASE test you are preparing to take and the educational tests you may be accustomed to.

Who Writes the Questions?

The questions, written by service industry experts familiar with all aspects of service consulting, are entirely job related. They are designed to test the skills that you need to know to work as a successful parts specialist; theoretical knowledge is not covered.

Each question has its roots in an ASE "item-writing" workshop where service representatives from automobile manufacturers (domestic and import), aftermarket parts and equipment manufacturers, working parts specialists, and vocational educators meet in a workshop setting to share ideas and translate them into test questions. Each test question written by these experts must survive review by all members of the group. The questions are written to deal with practical application of soft skills and product knowledge experienced by parts specialists in their day-to-day work.

All questions are pretested and quality-checked on a national sample of parts specialists. Those questions that meet ASE standards of quality and accuracy are included in the scored sections of the tests; the "rejects" are sent back to the drawing board or discarded altogether.

Each certification test is made up of between forty and eighty multiple-choice questions. The testing sessions are 4 hours and 15 minutes, allowing plenty of time to complete several tests.

Note: Each test could contain additional questions that are included for statistical research purposes only. Your answers to these questions will not affect your score, but since you do not know which ones they are, you should answer all questions in the test. The five-year Recertification Test will cover the same content areas as those listed above. However, the number of questions in each content area of the Recertification Test will be reduced by about one-half.

Objective Tests

A test is called an objective test if the same standards and conditions apply to everyone taking the test and there is only one correct answer to each question. Objective tests primarily measure your ability to recall information. A well-designed objective test can also test your ability to understand, analyze, interpret, and apply your knowledge. Objective tests include true-false, multiple choice, fill in the blank, and matching questions. ASE's tests consist exclusively of four-part multiple-choice objective questions.

Before beginning to take an objective test, quickly look over the test to determine the number of questions, but do not try to read through all of the questions. In an ASE test, there are usually between forty and eighty questions, depending on the subject. Read through each question before marking your answer. Answer the questions in the order they appear on the test. Leave the questions blank that you are not sure of and move on to the next question. You can return to those unanswered questions after you have finished the others. They may be easier to answer at a later time after your mind has had additional time to consider them on a subconscious level. In addition, you might find information in other questions that will help you to answer some of them.

Do not be obsessed by the apparent pattern of responses. For example, do not be influenced by a pattern like **D, C, B, A, D, C, B, A** on an ASE test.

There is also a lot of folk wisdom about taking objective tests. For example, there are those who would advise you to avoid response options that use certain words such as *all, none, always, never, must,* and *only,* to name a few. This, they claim, is because nothing in life is exclusive. They would advise you to choose response options that use words that allow for some exception, such as *sometimes, frequently, rarely, often, usually, seldom,* and *normally.* They would also advise you to avoid the first and last option (A and D) because test writers, they feel, are more comfortable if they put the correct answer in the middle (B and C) of the choices. Another recommendation often offered is to select the option that is either shorter or longer than the other three choices because it is more likely to be correct. Some would advise you to never change an answer since your first intuition is usually correct.

Although there may be a grain of truth in this folk wisdom, ASE test writers try to avoid them and so should you. There are just as many **A** answers as there are **B** answers, just as many **D** answers as **C** answers. As a matter of fact, ASE tries to balance the answers at about 25 percent per choice **A, B, C,** and **D.** There is no intention to use "tricky" words, such as outlined above. Put no credence in the opposing words "sometimes" and "never," for example.

Multiple-choice tests are sometimes challenging because there are often several choices that may seem possible, and it may be difficult to decide on the correct choice. The best strategy, in this case, is to first determine the correct answer before looking at the options. If you see the answer you decided on, you should still examine the options to make sure that none seem more correct than yours. If you do not know or are not sure of the answer, read each option very carefully and try to eliminate those options that you know to be wrong. That way, you can often arrive at the correct choice through a process of elimination.

If you have gone through all of the test and you still do not know the answer to some of the questions, then guess. Yes, guess. You then have at least a 25 percent chance of being correct. If you leave the question blank, you have no chance. In ASE tests, there is no penalty for being wrong.

Preparing for the Exam

The main reason we have included so many sample and practice questions in this guide is, simply, to help you learn what you know and what you don't know. We recommend that you work your way through each question in this book. Before doing this, carefully look through Section 3; it contains a description and explanation of the questions you'll find in an ASE exam.

Once you know what the questions will look like, move to the sample test. After you have answered one of the sample questions (Section 5), read the explanation (Section 7) to the answer for that question. If you don't feel you understand the reasoning for the correct answer, go back and read the overview (Section 4) for the task that is related to that question. If you still don't feel you have a solid understanding of the material, identify a good source of information on the topic, such as a textbook, and do some more studying.

After you have completed the sample test, move to the additional questions (Section 6). This time answer the questions as if you were taking an actual test. Once you have answered all of the questions, grade your results using the answer key in Section 7. For every question that you gave a wrong answer to, study the explanations to the answers and/or the overview of the related task areas.

Here are some basic guidelines to follow while preparing for the exam:

- Focus your studies on those areas you are weak in.
- Be honest with yourself while determining if you understand something.
- Study often but in short periods of time.
- Remove yourself from all distractions while studying.
- Keep in mind the goal of studying is not just to pass the exam, the real goal is to learn!

During the Test

Mark your bubble sheet clearly and accurately. One of the biggest problems an adult faces in test taking, it seems, is in placing an answer in the correct spot on a bubble sheet. Make certain that you mark your answer for, say, question 21, in the space on the bubble sheet designated for the answer for question 21. A correct response in the wrong bubble will probably be wrong. Remember, the answer sheet is machine scored and can only "read" what you have bubbled in. Also, do not bubble in two answers for the same question.

If you finish answering all of the questions on a test ahead of time, go back and review the answers of those questions that you were not sure of. You can often catch careless errors by using the remaining time to review your answers.

At practically every test, some technicians will invariably finish ahead of time and turn their papers in long before the final call. Do not let them distract or intimidate you. Either they knew too little and could not finish the test, or they were very self-confident and thought they knew it all. Perhaps they were trying to impress the proctor or other technicians about how much they know. Often you may hear them later talking about the information they knew all the while but forgot to respond on their answer sheet.

It is not wise to use less than the total amount of time that you are allotted for a test. If there are any doubts, take the time for review. Any product can usually be made better with some additional effort. A test is no exception. It is not necessary to turn in your test paper until you are told to do so.

Your Test Results!

You can gain a better perspective about tests if you know and understand how they are scored. ASE's tests are scored by American College Testing (ACT), a nonpartial, unbiased organization having no vested interest in ASE or in the automotive industry. Each question carries the same weight as any other question. For example, if there are fifty questions, each is worth 2 percent of the total score. The passing grade is 70 percent. That means you must correctly answer thirty-five of the fifty questions to pass the test.

The test results can tell you:

- where your knowledge equals or exceeds that needed for competent performance, or
- where you might need more preparation.

The test results *cannot* tell you:

- how you compare with other technicians, or
- how many questions you answered correctly.

Your ASE test score report will show the number of correct answers you got in each of the content areas. These numbers provide information about your performance in each area of the test. However, because there may be a different number of questions in each area of the test, a high percentage of correct answers in an area with few questions may not offset a low percentage in an area with many questions.

It may be noted that one does not "fail" an ASE test. The technician who does not pass is simply told "More Preparation Needed." Though large differences in percentages may indicate problem areas, it is important to consider how many questions were asked in each area. Since each test evaluates all phases of the work involved in a service specialty, you should be prepared in each area. A low score in one area could keep you from passing an entire test.

There is no such thing as average. You cannot determine your overall test score by adding the percentages given for each task area and dividing by the number of areas. It doesn't work that way because there generally are not the same number of questions in each task area. A task area with twenty questions, for example, counts more toward your total score than a task area with ten questions.

Your test report should give you a good picture of your results and a better understanding of your task areas of strength and weakness.

If you fail to pass the test, you may take it again at any time it is scheduled to be administered. You are the only one who will receive your test score. Test scores will not be given over the telephone by ASE nor will they be released to anyone without your written permission.

3 Types of Questions on an ASE Exam

ASE certification tests are often thought of as being tricky. They may seem to be tricky if you do not completely understand what is being asked. The following examples will help you recognize certain types of ASE questions and avoid common errors.

Each test is made up of forty to eighty multiple-choice questions. Multiple-choice questions are an efficient way to test knowledge. To answer them correctly, you must think about each choice as a possibility, and then choose the one that best answers the question. To do this, read each word of the question carefully. Do not assume you know what the question is about until you have finished reading it.

About 10 percent of the questions on an actual ASE exam will use an illustration. These drawings contain the information needed to correctly answer the question. The illustration must be studied carefully before attempting to answer the question. Often, techs look at the possible answers then try to match up the answers with the drawing. Always do the opposite; match the drawing to the answers. When the illustration is showing an electrical schematic or another system in detail, look over the system and try to figure out how the system works before you look at the question and the possible answers.

Multiple-Choice Questions

The most common type of question used on ASE Tests is the multiple-choice test. This type of question contains 3 "distracters" (wrong answers) and one "key" (correct answer). When the questions are written effort is made to make the distracters plausible to draw an inexperienced technician to one of them. This type of question gives a clear indication of the technician's knowledge. Using multiple criteria including cross-sections by age, race, and other background information, ASE is able to guarantee that a question does not bias for or against any particular group. A question that shows bias toward any particular group is discarded. If you encounter a question that you are unsure of, reverse engineer it by eliminating the items that it cannot be. For example:

A rocker panel is a structural member of which vehicle construction type?

A. Front-wheel drive
B. Pickup truck
C. Unibody
D. Full-frame

Analysis:

This question asks for a specific answer. By carefully reading the question, you will find that it asks for a construction type that uses the rocker panel as a structural part of the vehicle.

Answer A is wrong. Front-wheel drive is not a vehicle construction type.
Answer B is wrong. A pickup truck is not a type of vehicle construction.
Answer C is correct. Unibody design creates structural integrity by welding parts together, such as the rocker panels, but does not require exterior cosmetic panels installed for full strength.
Answer D is wrong. Full-frame describes a body-over-frame construction type that relies on the frame assembly for structural integrity.

Therefore, the correct answer is C. If the question was read quickly and the words "construction type" were passed over, answer A may have been selected.

EXCEPT Questions

Another type of question used on ASE tests has answers that are all correct except one. The correct answer for this type of question is the answer that is wrong. The word **"EXCEPT"** will always be in capital letters. You must identify which of the choices is the wrong answer. If you read quickly through the question, you may overlook what the question is asking and answer the question with the first correct statement. This will make your answer wrong. An example of this type of question and the analysis is as follows:

All of the following are tools for the analysis of structural damage **EXCEPT:**

A. height gauge
B. tape measure.
C. dial indicator.
D. tram gauge.

Analysis:

The question really requires you to identify the tool that is not used for analyzing structural damage. All tools given in the choices are used for analyzing structural damage except one. This question presents two basic problems for the test-taker who reads through the question too quickly. It may be possible to read over the word **"EXCEPT"** in the question or not think about which type of damage analysis would use answer C. In either case, the correct answer may not be selected. To correctly answer this question, you should know what tools are used for the analysis of structural damage. If you cannot immediately recognize the incorrect tool, you should be able to identify it by analyzing the other choices.

Answer A is wrong. A height gauge *may* be used to analyze structural damage.
Answer B is wrong. A tape measure may be used to analyze structural damage.
Answer C is correct. A dial indicator may be used as a damage analysis tool for moving parts, such as wheels, wheel hubs, and axle shafts, but would not be used to measure structural damage.
Answer D is wrong. A tram gauge *is* used to measure structural damage.

Technician A, Technician B Questions

The type of question that is most popularly associated with an ASE test is the "Technician A says . . . Technician B says . . . Who is right?" type. In this type of question, you must identify the correct statement or statements. To answer this type of question correctly, you must carefully read each technician's statement and judge it on its own merit to determine if the statement is true.

Typically, this type of question begins with a statement about some analysis or repair procedure. This is followed by two statements about the cause of the problem, proper inspection, identification, or repair choices. You are asked whether the first statement, the second statement, both statements, or neither statement is correct. Analyzing this type of question is a little easier than the other types because there are only two ideas to consider although there are still four choices for an answer.

Technician A, Technician B questions are really double true or false questions. The best way to analyze this kind of question is to consider each technician's statement separately. Ask yourself, is A true or false? Is B true or false? Then select your answer from the four choices. An important point to remember is that an ASE Technician A, Technician B question will never have Technician A and B directly disagreeing with each other. That is why you must evaluate each statement independently. An example of this type of question and the analysis of it follows.

Structural dimensions are being measured. Technician A says comparing measurements from one side to the other is enough to determine the damage. Technician B says a tram gauge can be used when a tape measure cannot measure in a straight line from point to point. Who is right?

A. A only
B. B only
C. Both A and B
D. Neither A nor B

Analysis:

With some vehicles built asymmetrically, side-to-side measurements are not always equal. The manufacturer's specifications need to be verified with a dimension chart before reaching any conclusions about the structural damage.

Answer A is wrong. Technician A's statement is wrong. A tram gauge would provide a point-to-point measurement when a part, such as a strut tower or air cleaner, interrupts a direct line between the points.
Answer B is correct. Technician B is correct. A tram gauge can be used when a tape measure cannot be used to measure in a straight line from point to point.
Answer C is wrong. Since Technician A is not correct, C cannot be the correct answer.
Answer D is wrong. Since Technician B is correct, D cannot be the correct answer.

Most-Likely Questions

Most-Likely questions are somewhat difficult because only one choice is correct while the other three choices are nearly correct. An example of a Most-Likely-cause question is as follows:

The Most-Likely cause of reduced turbocharger boost pressure may be a:

A. wastegate valve stuck closed.
B. wastegate valve stuck open.
C. leaking wastegate diaphragm.
D. disconnected wastegate linkage.

Analysis:

Answer A is wrong. A wastegate valve stuck closed increases turbocharger boost pressure.
Answer B is correct. A wastegate valve stuck open decreases turbocharger boost pressure.
Answer C is wrong. A leaking wastegate valve diaphragm increases turbocharger boost pressure.
Answer D is wrong. A disconnected wastegate valve linkage will increase turbocharger boost pressure.

LEAST-Likely Questions

Notice that in Most-Likely questions the first word is not all capitalized. This is not so with LEAST-Likely type questions. For this type of question, look for the choice that would be the LEAST-Likely cause of the described situation. Read the entire question carefully before choosing your answer. An example is as follows:

What is the LEAST-Likely cause of a bent pushrod?

- A. Excessive engine speed
- B. A sticking valve
- C. Excessive valve guide clearance
- D. A worn rocker arm stud

Analysis:

Answer A is wrong. Excessive engine speed may cause a bent pushrod.
Answer B is wrong. A sticking valve may cause a bent pushrod.
Answer C is correct. Excessive valve clearance will not generally cause a bent pushrod.
Answer D is wrong. A worn rocker arm stud may cause a bent pushrod.

Summary

There are no four-part multiple-choice ASE questions having "none of the above" or "all of the above" choices. ASE does not use other types of questions, such as fill-in-the-blank, completion, true-false, word-matching, or essay. ASE does not require you to draw diagrams or sketches. If a formula or chart is required to answer a question, it is provided for you. There are no ASE questions that require you to use a pocket calculator.

Testing Time Length

An ASE test session is four hours and fifteen minutes. You may attempt from one to a maximum of four tests in one session. It is recommended, however, that no more than a total of 225 questions be attempted at any test session. This will allow for just over one minute for each question.

Visitors are not permitted at any time. If you wish to leave the test room, for any reason, you must first ask permission. If you finish your test early and wish to leave, you are permitted to do so only during specified dismissal periods.

You should monitor your progress and set an arbitrary limit to how much time you will need for each question. This should be based on the number of questions you are attempting. It is suggested that you wear a watch because some facilities may not have a clock visible to all areas of the room.

4 Overview of the Task List

Automobile Parts Specialist (Test P2)

The following section includes the task areas and task lists for this test and a written overview of the topics covered in the test.

The task list describes the actual work you should be able to do as a technician that you will be tested on by the ASE. This is your key to the test and you should review this section carefully. We have based our sample test and additional questions upon these tasks, and the overview section will also support your understanding of the task list. ASE advises that the questions on the test may not equal the number of tasks listed; the task lists tell you what ASE expects you to know how to do and be ready to be tested upon.

At the end of each question in the Sample Test and Additional Test Questions sections, a letter and number will be used as a reference back to this section for additional study. Note the following example: **C.14.10**.

Task List

C. Vehicle Systems Knowledge (36 Questions)

14. Miscellaneous (2 Questions)

Task C.14.10 — Recommend proper application and usage of chemicals.

Example:

119. Materials that become unstable and are likely to burn, explode, or give off toxic fumes are called:
    A. flammable materials.
    B. corrosive materials.
    C. reactive materials.
    D. toxic materials.

(C.14.10)

Analysis:

Question #119
Answer A is wrong. Flammable materials are those that easily catch fire or explode.
Answer B is wrong. Corrosive materials are those that will dissolve metal.
Answer C is correct. Reactive materials are those that become unstable and are likely to burn, explode, or give off toxic fumes.
Answer D is wrong. Toxic materials cause injury or death from contact, ingestion, or inhalation.

Task List and Overview

A. General Operations (10 Questions)

Task A.1 **Calculate discounts, profits, percentages, and pro-rated warranties.**

  a. To calculate percent discount:
    1. Convert the percent into a decimal number by moving the decimal two places to the left. For example, 10 percent (10%) becomes 0.10.
    2. Multiply the decimal number by the original price to get the amount of the discount. For example, 10% of $25.00 is $2.50 (0.10 × $25.00 = $2.50).
  b. To calculate percent of profit:
    1. Subtract the cost from the selling price. For example, if the selling price is $15.00 and the cost is $10.00, the profit is $5.00. ($15.00 − $10.00 = $5.00).
    2. Divide the profit by the selling price; $5.00 divided by $15.00 is 0.33.
    3. Convert this number into a percentage by moving the decimal two places to the right; 0.33 becomes 33 percent (33%). The profit is 33%.

Task A.2 **Calculate special handling charges.**

The customer received $95.00, which is 95% of $100.00. The restock fee, usually expressed as a percentage, is the fee charged for having to handle a returned part.
  a. To calculate the restocking fee, consider 5 percent:
    1. Convert the percent into a decimal number by moving the decimal two places to the left; 5 percent becomes 0.05.
    2. Multiply the decimal number by the original price to get the amount of the restocking fee; 0.05 × $100.00 is $5.00.
    3. Subtract the restocking fee ($5.00) from the original price ($100.00) to get the amount of money to be returned to the customer. $100.00 − $5.00 = $95.00.

Task A.3 **Identify and convert units of measure.**

If a customer came into your store and needed 5 liters of oil, how many quarts would you give him? One quart is equal to 0.9464 liters, so one quart is almost equal to one liter. Therefore, 5 liters is approximately equal to 5 quarts. In actual practice, this amount will leave the engine ¼ quart low on oil but within the safe operating range.

There are 61 cubic inches to a liter. To convert to cubic inches multiply the liters by 61 to get cubic inches. For example, a 5 liter engine is the same as a 305 cubic inch engine (5 × 61 = 305). The metric unit of temperature measurement is degrees Celsius (° C). To convert from degrees Fahrenheit (° F) to degrees Celsius using a short formula, first subtract 32 then multiply by 0.555. For example, 212° F is equal to 99.9° C (212 − 32 = 180, and 180 × 0.555 = 99.9). Using a long formula, which yields a more accurate conversion, 212° F is equal to 100° C.

Task A.4 **Determine alphanumeric sequences.**

An alphanumeric listing places a series of numbers and letters in order starting from the left digit and working across to the right. If the numbers are the same, the order sequence continues with the letter A, then B, and so forth.

Task A.5 **Determine sizes with precision measuring tools and equipment.**

Take for example the figure reading for question #5 in the Sample Test, 0.184 inch. Because this is a 0–1 inch micrometer, the reading must be between 0 and 1. Since the 1 is the last complete unit visible on the horizontal line, the first measurement is 0.100. Three ¹⁄₄₀-inch marks are visible after the 1, so the second measurement is 3 × 0.025 = 0.075. The horizontal line nearly lines up with the 9 on the vertical line, so 0.009 is the third measurement. Adding up the three measurements (0.100 + 0.075 + 0.009) gives a total reading of 0.184.

Task A.6 Perform money transactions (cash, checks, credit and debit cards).

The customer should always pay the amount that is on the total invoice line.

Cash sales are often the most profitable for a business; little paperwork is necessary and, usually, little time is required. With a cash sale, cash flow is not impeded. Offering discounts for cash purchases and making sure cash customers are handled courteously and efficiently encourage customers to pay with cash. The following are a few points for conducting a cash sale effectively:

a. Greet the customer and determine his or her needs. Write this information down to ensure accuracy.
b. Refer to catalogs, note the necessary information, and check inventory.
c. Pull the merchandise and suggest related items and/or services that can benefit the customer.
d. Fill out an invoice and include any applicable discounts.
e. Accept payment for the purchase and make change, if necessary.
f. Thank the customer, remind him or her of the warranty policy, and invite him or her to return.

The process for performing sales with other forms of payment differ only in the method of payment and some procedures that are added to the transaction. The following additional items relate to checks:

a. Verifying the customers information on the check.
b. Checking information against an i.d. and adding that information to the check.
c. Some businesses collect a second form of i.d. by way of a credit card.
d. Running the check through a check verification process either electronic or by telephone.
e. Applying endorsement to the check before putting it in the cash drawer.
f. Posting the check number in the Point of Sale system.

Your business may use some or none of these processes but, they are good to be aware of.

Credit cards and debit cards work almost the same as one another. Here are some tasks you may need to perform when working with them:

a. Verifying the card belongs to the customer.
b. Checking information against an i.d.
c. Many credit card processors offer small flat fees if your customer enters a PIN number and pays as a debit, so it is good to spot those cards that say "check" or "debit" and ask if the customer would be willing to enter their PIN. This will save your business money on transaction fees.
d. If a sale needs to be voided or credited you should know the steps necessary to perform the functions. In general, if the transaction took place that day you can void it and do a new transaction; if the system has already performed batching you will have to do a credit. These will be specific to your business and your processor.

Task A.7 Perform sales and credit invoicing.

Charge account sales encourage large purchases or purchases from customers who do not want to carry cash or do not have the cash. Counter personnel must be familiar with the charge policy in their workplace to ensure accuracy and customer satisfaction.

Task A.8 Interact with management and fellow employees.

Parts specialists interact with people. They work with customers, suppliers, manufacturers, and numerous other segments of the aftermarket industry. The ability to communicate is essential, whether it is explaining the features and benefits of a product to a customer, placing an order with a warehouse distributor, or routing the proper paperwork and data through the various departments in the jobber store.

Interacting with others is not always easy. When personalities clash, misunderstandings will occur. A parts specialist or manager must be a problem solver, an astute negotiator, and a person who gains satisfaction through serving and helping others.

Task A.9 Understand the value of housekeeping skills (facility, work stations, and backroom).

It is the responsibility of all employees to keep all areas of the store clean and orderly. All employees suffer from lost sales if the floors, shelves, and displays are a safety hazard or do not appeal to the customer.

Task A.10 Assist with employee and customer training.

The parts specialist should seek whatever help is necessary in order to sell the correct parts and satisfy the customer. In addition, experienced parts specialists are often asked to help train new employees.

Task A.11 Identify potential safety risks; demonstrate proper safety practices.

Shipping, receiving, and stocking materials requires physical exertion. Knowing the proper way to lift heavy materials is important. Always lift and work within your ability and seek help from others when you are not sure if you can handle the size or weight of the material or object. Auto parts, even small, compact components, can be surprisingly heavy or unbalanced. Always size up the lifting task before beginning.

Task A.12 Identify proper handling of regulated and/or hazardous materials.

The Environmental Protection Agency (EPA), Occupational Safety and Health Administration (OSHA), and other state and local agencies have strict guidelines for handling these materials. OSHA's Hazard Communication Standard (HCS), commonly called the "Right-to-Know Law," applies to all companies that use or store any kind of hazardous chemicals that workers might come in contact with, including: solvents, caustic cleaning compounds, abrasives, cutting oils, and other hazardous materials.

Compliance with the law requires a system for labeling hazardous chemicals and maintaining Material Safety Data Sheets (MSDS). These sheets must be made available to employees, informing them of the dangers inherent in chemicals found in the workplace.

Task A.13 Identify potential security risks.

Expensive items should be displayed behind the counter in locked cases. Parts specialists should give a lot of thought to what gets displayed near entrances and exits. In the moment that a parts specialist is busy researching parts in a catalog, a thief can easily slip something under his or her jacket and glide through the door. It is best to keep the areas around the entrances clear, or to display only large, heavy, or awkward merchandise near the door. Barrels of oil and bags of floor sweep are safe choices for displays near entrances.

Task A.14 Identify parts industry terminology.

An invoice should be completed for each sale in the store. These invoices are used to track inventory and customer purchases for billing purposes. A purchase order is used to allow a company to purchase parts. It will describe the parts and quantity of parts to be purchased along with billing information. A stock order is used by the store to order more stock from the suppliers. A back order refers to merchandise ordered from a supplier but not shipped, due to the supplier being out of stock. An emergency order is an order placed with the supplier on a routine basis, usually weekly or biweekly.

Task A.15 Understand the value of company policies and procedures.

Company policies and procedures define how certain things should be handled. They are also the guidelines for making decisions. The policies will dictate who gets how much of a discount. They will define the return policy and other customer service routines. Procedures define how things need to be reported and recorded, as well as the steps to be taken when something unusual comes up. By adhering to and understanding the company's policies and procedures, a parts specialist will be able to offer consistent customer service.

Task A.16 Understanding the basic functions of tools and equipment used in automotive service.

The most basic understanding of tools and equipment include a discussion of hand tools and mechanical tools. The tools most commonly found in the industry fall into these two categories. Hand tools include, but are not limited to the following. There are many types of wrenches (which means to twist), including the open end, box end combination, socket, and adjustable wrenches. Each of these is used to tighten or loosen bolts or fasteners. Screwdrivers are defined by their size: standard tip, Phillips, pozidriv, torx, clutch, or scrulox (which tighten or loosen fasteners as well). Pliers are gripping tools and come in several styles: combination, adjustable, needle nose, locking, vise grips, diagonal cutting, snap, and retaining ring. Hammers are defined by the material and weight of the head: steel or soft faced. Tap and Dies are used to cut threads. The tap cuts threads internally and the die cuts them on the outside of the bolt. Major mechanical equipment includes the hydraulic lift, the pneumatic air gun, a floor mounted hydraulic press, and possibly the computer to operate a repair facility.

B. Customer Relations and Sales Skills (11 Questions)

Task B.1 Identify customer types and skill level.

Do-it-yourselfers usually need more information than do professional automotive customers. These customers usually need and expect good advice on parts and methods. They view the parts specialist as an expert, and the parts specialist should utilize product knowledge and catalog skills to give the advice needed. Nothing should be sold to a customer that is not actually needed.

One way that a parts specialist can assist do-it-yourselfers without taking too much time away from other customers is to have printed how-to information available. Once the parts specialist determines what work the customer will be doing, the parts specialist can provide something to read while assisting other customers. After the customer reads the information, the parts specialist can clarify some important points and answer questions. Each customer's needs can be met without anyone waiting a long time.

On the other hand a professional technician might consider it bad form if you offer "advice". Your job is to determine the point where your incredibly valuable experience or information about a special consideration when using a given product crosses over the line to telling your customer that you think he might not have the skills to perform his job.

Task B.2 Identify customer needs.

A parts specialist's opening question should be designed to put the customer at ease and elicit as much data as possible. The question, "May I help you?" is the worst possible opening because it invites the opportunity for the customer to reply, "No." Instead, the parts specialist should ask, "How can I help you?" Another good opening is, "Can I show you anything in particular or would you like to browse around first?"

These questions restrict answers to the positive and give the customer a way out without having to say, "No." The customer's response will generally indicate whether he or she has a defined objective, is just looking, or is seriously considering a purchase.

Task B.3 Provide information.

Some customers simply lack confidence rather than skill. Providing information can boost confidence. The parts specialist's greatest impact on customers' confidence is his or her own attitude about their abilities. Avoid speaking down to customers, and converse with them according to their actual abilities. Some problems may be beyond the capabilities of some do-it-yourselfers.

Task B.4 Handle customer complaints and returns.

The parts specialist must not be concerned with placing blame because it is not always evident whether or not there is any blame to place. The customer is not always right, but the parts specialist must be very careful when making corrections. Embarrassed customers will probably spend their money somewhere else.

Often the easiest way for the parts specialist to deal with a difficult situation is to put themselves in the customer's shoes. Most people, including parts counter personnel, are angry when something that was purchased causes a problem. Often the anger is directed toward the person who sold it. The parts specialist should take whatever time is necessary to listen to and try to understand the customer's viewpoint.

Task B.5 Acknowledge/greet the customer.

From the time a customer walks through the door until the time he or she leaves, that person should be acknowledged and responded to favorably. Counter personnel should refrain from personal conversations with one another whenever a customer approaches the counter.

Customers usually do not want to feel like a number or just another face in the crowd. The first step toward making a customer feel noticed is to make eye contact with the customer by looking at him or her and saying something such as, "Good morning. I'll be with you in a moment." Eye contact should also be made when the parts specialist is waiting on the customer by looking up from the parts catalog periodically while talking to him or her.

Task B.6 Demonstrate proper telephone skills.

Often the phone will ring while the parts specialist is serving a customer. If no one else is available to answer the phone, the parts specialist should politely ask the customer to wait and then answer the phone. If it appears that the call will take a while, the parts specialist should explain to the caller that he or she is assisting another customer at the moment. After taking pertinent information, the parts specialist should state that he or she will return the call as soon as possible or that the caller should phone back in a few minutes. That way, neither the caller nor the counter customer is kept waiting very long.

Since a caller cannot see the parts specialist's facial expressions, gestures, or body language, all communications must be accomplished with words, tones, timing, and inflection. Parts counter personnel must remember to speak clearly and slowly enough to be understood and must request that the customer do so if not clearly understood.

Task B.7 Obtain pertinent application information.

Defensive selling implies a method of selling that protects the interests of the store in response to do-it-yourselfers who are not always clear and methodical in their diagnosis of a problem. These customers will sometimes purchase the wrong part, install it, find that the problem has not been solved, and then try to return the part as defective. Parts that are sold and returned in this manner are so done at the expense of the store. For this reason, many stores have a policy that does not allow the return of parts that have been installed.

Task B.8 Present a knowledgeable and professional image.

The appearance of the store, as well as the appearance of counter personnel, influences the customer. A clean, neat parts specialist establishes a positive image with customers, and a clean, neat store is equally important.

Although a parts specialist is rarely in charge of decor or layout, he or she can maintain the appearance of the store in several ways:

a. Keep a rag handy for wiping the counter so that it is free of dirt and grime from used parts.
b. Keep the counter free of clutter such as small parts, notes, paper clips, and other such items.
c. Keep the displays organized and well stocked.
d. Keep the stock and supplies neatly shelved to improve appearance.
e. Assist with basic clean-up by throwing away empty parts wrappers and labels and by removing empty boxes from sight.

Task B.9 Recognize the value of selling related items.

Selling related parts when a customer buys a particular item can boost profits by 30 percent or more. Selling related parts also makes sense when the parts specialist remembers that a customer's problem might not be solved solely by replacing a faulty part if the related hardware or chemicals are not also up to peak performance. For example, if selling a wheel cylinder, suggest brake fluid.

If the parts specialist suggests related items on the mechanic's first visit, both the customer and the parts specialist can save a lot of time. The customer will be very appreciative of the parts specialist's initial time investment, and with this type of service will more than likely return for parts again.

Task B.10 Identify product features and benefits.

When the parts specialist carries many different brands of the same product, the difference between them should be explained. If the store carries, for example, four different kinds of brake pads, the parts specialist should identify the material, size, cost, warranty, and name brand differences. The parts specialist should let the customer decide on which part to buy. The parts specialist can make recommendations but the final decision should be the responsibility of the customer.

Task B.11 Handle objections.

After some experience, a parts specialist will know when a customer is ready to buy and when a customer is there to browse. A parts specialist should not push the sale of any parts that a customer may not need. Part of a parts specialist's job is to be able to identify the different customer types and determine their needs.

Task B.12 Balance telephone and in-store customers.

If a parts specialist must put a telephone customer on hold, he or she should identify the name of the store and politely state something such as, "Please hold one moment." The parts specialist should never pick up the phone and press the hold button without saying anything, and he or she should never push the button before finishing a sentence or phrase. If the parts specialist knows that it will take a while before he or she can talk with the caller, the parts specialist should ask that the caller call back rather than be put on hold for a long period of time.

Task B.13 Promote store services and features.

A good parts specialist should always promote sales and store services. If a customer asks for some brake pads, the parts specialist should let the customer know that they have a machine shop and can turn his rotors or drums if needed. The parts specialist can also let the customer know that they have brake fluid and other related brake parts in stock to further promote the store sales.

Task B.14 Promote upgraded products.

Selling related items is extremely important, not only to build profits, but also to help the customer achieve the safest possible results. Displays should be set up to advertise the complete service job. Brakes pads, shoes, and hardware once came in fairly plain, dull boxes, but today, most manufacturers have upgraded their packaging graphics to the point where these products can be used to build attractive, effective displays in the front of the store.

Task B.15 Solve customer problems.

A parts specialist should help the customer get the right part each and every time. If a customer comes in for a starter while carrying a pair of jumper cables, the parts specialist might ask, "What was your car's problem that lead you to believe that it needs a starter?" The customer may say, "It will not start without jump-starting the battery." The parts specialist should then explain that the battery could be the problem. This way the parts specialist will not only get the sale for the battery but will not have sold the customer a part that did not fix the vehicle.

Task B.16 Close sale.

Closing simply means getting a commitment from the customer. Successful salespersons usually share certain characteristics. First, effective closers expect the sale. The very best closers are certain that they will bring each interview to a successful close. Their confidence might be justified, or it might be the product of an inflated ego, but, whatever the source, that expectation of success often results in sales. Second, good closers always let the customer know that they expect the sale. Finally, many parts specialists do a fantastic job of qualifying prospects, matching benefits to needs, laying a foundation of agreement, handling any objections, and then stopping and waiting for customers to start shelling out their money.

C. Vehicle Systems Knowledge (40 Questions)

1. Engine Mechanical Parts (3 Questions)

Tasks
C.1.1
C.1.2
C.1.3

Identify major and related components and their function.

For Tasks C.1.1 through C.1.3, we are providing you with an itemized list of the major components in a format familiar to you as a parts specialist. Each item in the diagram has a matching description in the table that follows.

Overview of the Task List   Automotive Parts Specialist (Test P2)

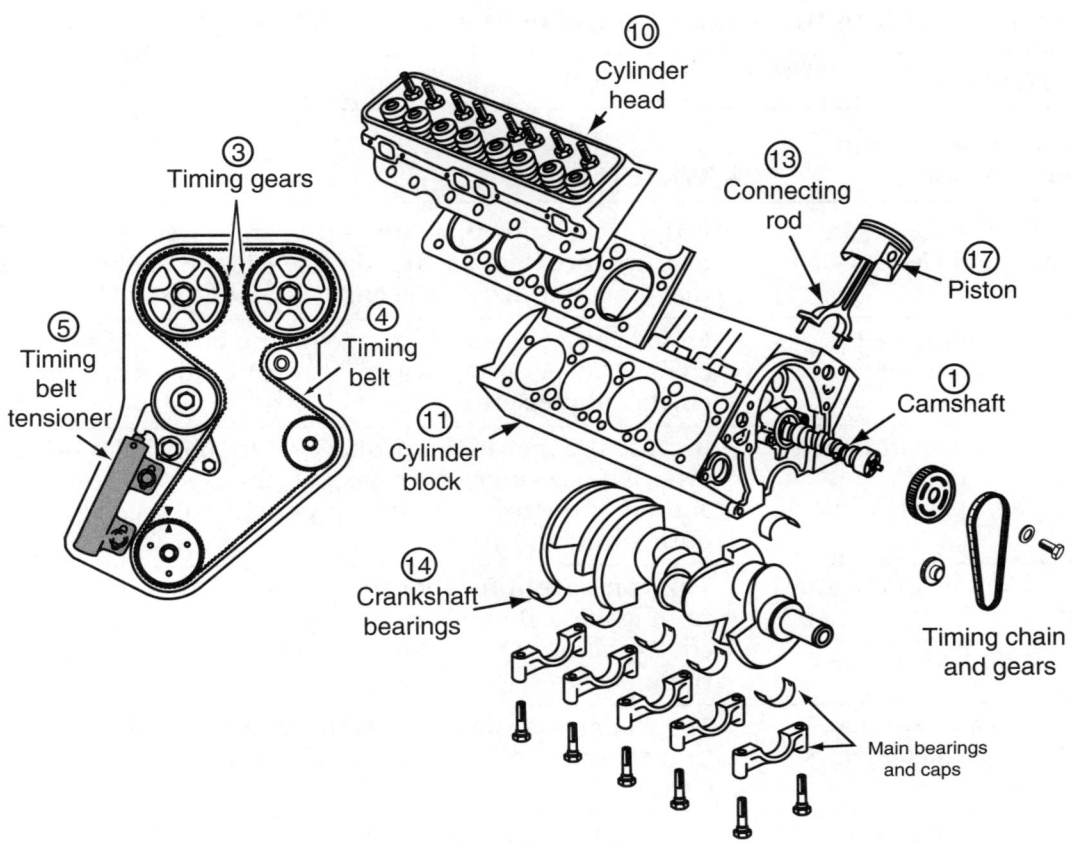

Exploded View—Engine

VALVE TRAIN

Diagram Number	Component Name	What It Does
1.	Camshaft	The camshaft has eccentric lobes that rise off the centerline of the shaft to push on valve lifters or rocker arms and cause valves to open. The cam is driven at half-crank speed by timing chain, timing belt, or timing gears connected to the crankshaft.
2.	Valves	Intake valves control the introduction of fuel and air into the engine and open at the beginning of the intake stroke. Exhaust valves control the elimination of by-products of the combustion process. Both valves are closed during the compression and power strokes.
3.	Timing Gears	These are driven by the timing belt or a chain. The example shown is a belt drive design dual overhead camshaft.
4.	Timing Belt	Made of rubber with reinforcing bands and teeth made of rubber to mesh with cam and crank gears. Many engines have valve train that is known as interference. This means that if the timing belt breaks, the valves will come in contact with the pistons. This is why it is critical to follow the mileage intervals recommended for belt replacement. It is still important in non-interference engines (engines that will not have valve to piston contact if the belt breaks) to follow the intervals to avoid a break down and resulting tow to the shop.

EXPLODED VIEW—4-CYLINDER DOHC ENGINE—*Continued*

VALVE TRAIN

Diagram Number	Component Name	What It Does
Not Numbered	Timing Chain and Gears	Perform the same function as the timing belt. They are not a maintenance item like the timing belt and are internal components that are lubricated by the engine's oiling system.
5.	Timing Belt Tensioner	Provide proper pressure on the timing belt or, in some cases, timing chain. May be spring loaded, manually adjusted, or actuated by engine oil pressure.
Not Pictured	Camshaft Pulley Seals	Because the camshaft must have oil to lubricate it, when the cam drive is a belt drive, there are seals that keep the oil in the engine and off of the belt. This is a very commonly replaced component during timing belt service.
Not Pictured	Front Crank Seal	This component functions the same as a cam seal. You will find a front crank seal in all engines, regardless of valve train configuration. This component often fails and is replaced during timing belt service.
Not Pictured	Valve Springs and Retainers	These apply pressure to keep the opening and closing of the valve in matching motion to the lobes of the camshaft and to hold them closed.
Not Pictured	Timing Cover	Used to protect the timing belt. In timing chain applications, it covers the timing chain, controls oil, and often contains the oil pump.
Not Pictured	Valve Cover	The cover that attaches to the cylinder head and covers the valve train components.
10.	Cylinder Head	Holds all of the valve train components and has air flow ports for the exhaust and intake. The cylinder head bolts onto the cylinder block and has the intake and exhaust manifolds bolted to it. The complete package is often called the induction system.

SHORT BLOCK ASSEMBLY

11.	Cylinder Block	The backbone of the engine all major components bolt to it. The cylinder bores are the large holes in it. The pistons go in these holes. The crankshaft mounts in bearing insert in the blocks main saddles.
Not Numbered	Crank Shaft	The crankshaft is a large offset shaft. The journals are configured in 60, 90, or 180-degree relationships, depending on the type of engine. The large weights hanging opposite of the journals are used to counter-balance the weight of the journal and the rod and piston assembly that attaches to each journal. In most V6 or V8 engines, there are two connecting rods bolted Siamese to each crank journal, while 4-cylinder engines usually have one rod on each crank journal. This is a heavy component made of steel or nodular iron in most cases.
13.	Connecting Rod	The connecting rod has two holes in it when viewed by itself. The large hole is the side that holds the rod bearing and splits apart to bolt to the crank journal. The small end accepts the wrist pin of the piston, which is the point the piston pivots on when the crankshaft is spinning.

Overview of the Task List

EXPLODED VIEW—4-CYLINDER DOHC ENGINE—*Continued*

SHORT BLOCK ASSEMBLY

Diagram Number	Component Name	What It Does
14.	Crank Bearings	The bearings in modern engines are a composite of different materials clad together. The back shell is usually aluminum, with a copper layer on it and then a soft metal alloy similar to lead that is the actual bearing surface. The bearing does not have any moving parts like other bearings. In the engine, its job is to carry an oil film to the surfaces of the crank and maintain adequate clearance between moving parts. When the clearances are excessive, knocking noise and loss of oil pressure can result. If the bearing clearance is too tight, the oil film that cools and protects the parts is too thin to be effective, causing damage of the parts.
Not Pictured	Oil Pump	The heart of the engine. Oil pumps may be driven directly by the crankshaft or by gears and a driveshaft from the valve train. The pump pulls oil from the oil pan and passes it to the oil filter. Because it can move more volume than the internal clearances of the engine can flow, pressure is created in the oiling system. This pressure occurs in small passages inside the engine, similar to the plumbing inside a house, carrying oil to all of the bearing surfaces in the engine. The oil pan and valve covers are not under pressure. They simply keep the oil in the engine. The oil pan is a reservoir for the oil to return to. Engines use drains like the gutters on a house to return unpressurized oil back to the pan at the bottom of the engine.
Not Pictured	Rear Main Seal	This is the large seal that keep the oil from the rear crank main bearing inside the engine. This seal may be a 2-piece design that is serviced by removing the rear main bearing cap, or it may be a 1-piece full circle seal that is installed from the outside rear of the engine.
17.	Piston	The piston is made of aluminum in all modern engines. When it moves to the bottom of the cylinder, it creates a vacuum that pulls air and fuel into the cylinder; this is called the intake stroke. It then rises to the top to compress the air it just took in (called the compression stroke) before the spark plug sparks and causes the high-pressure fuel and air mixture to create lots of heat as it oxidizes (burns). This heat causes expansion, which forces the piston back down; this is the power stroke. This causes the crank to spin because of those eccentric journals discussed earlier. The crank continues to spin on inertia, resulting in final stroke in the 4-stroke cycle, known as the exhaust stroke. This is where we get rid of the left-over by-products of combustion. Since these cycles happen in the other cylinders at complimentary times, the crank is being pushed by a compression stroke at least every 90° of rotation.
Not Pictured	Piston Rings	There are two types of piston rings on each piston. One is the compression ring. There are usually two of these stacked in grooves in the piston, right on top of each other. Since the piston must be lubricated to slide up and down the cylinder, it cannot be too tight in the bore. The compression rings are responsible for providing an extremely effective and small sealing surface that requires very little lubrication. The second ring type is the oil ring which helps to carry oil up the cylinder bore to lubricate and to scrape it back off and pass it through notches cut in the piston skirt on the way back down.

Task C.1.4

Provide basic use and installation instructions.

Many of the products sold for the engine are offered in remanufactured, rebuilt, and new conditions. You will need to help your customers choose the product for their needs and provide them with information from the manufacturer of the part to help them with proper installation. Engines often come with a warranty that is dependent on other components and systems. Your customers can benefit from your knowledge of these warranty programs.

2. Cooling Systems (2 Questions)

Tasks C.2.1 C.2.2 C.2.3

Identify major and related components and their function.

The cooling system serves two different purposes. First, it keeps the engine at a steady temperature. Second, it provides the hot water that makes the heater work. The cooling system is made up of many components. The diagram below will show a basic system common to most vehicles. The cooling system contains a mixture of antifreeze and water. The mixture carries heat from the engine components to the radiator and the heater core where the heat is exchanged by convection between the outside air and the radiator or heater core. Let's take a look at the components and how they work together.

For Tasks C.2.1 through C.2.3, we are providing you with an itemized list of the major components in a format familiar to you as a parts specialist. Each item in the diagram has a matching description in the table that follows.

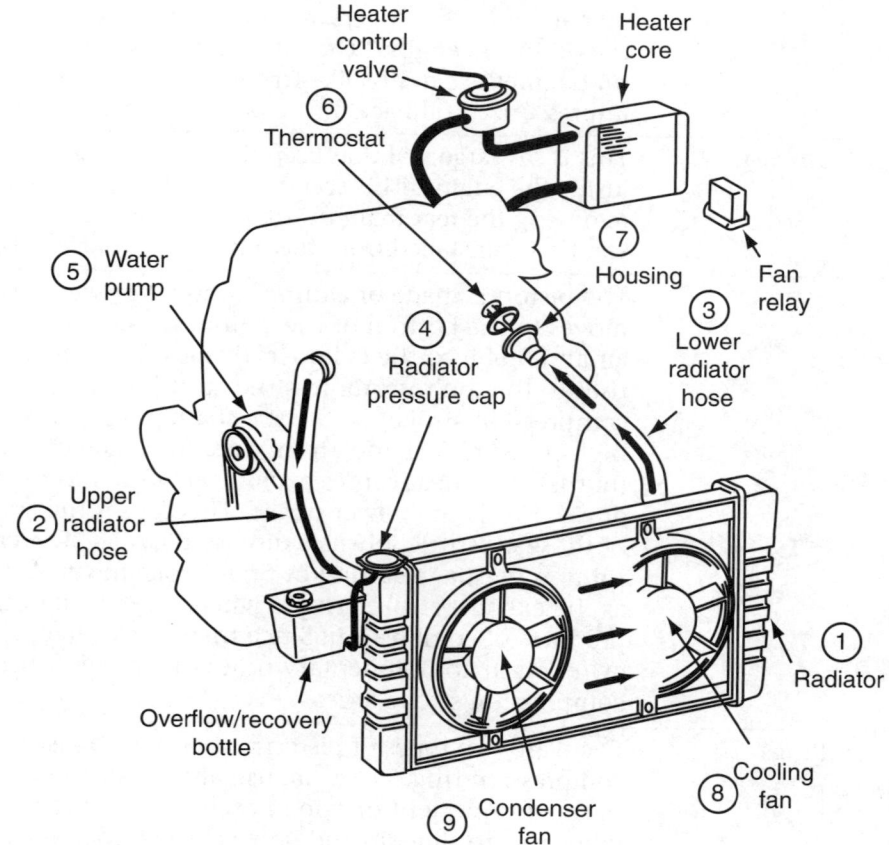

Cooling System Exploded View

COOLING SYSTEM EXPLODED VIEW

Diagram Number	Component Name	What It Does
1.	Radiator	At its very simplest, the radiator is a heat exchanger. Warm coolant enters it, the outside air moving across it removes heat from the coolant, and it returns to the engine to start the cycle over again. Radiators are carefully chosen to have the correct amount of heat transfer efficiency called BTU (British thermal units). If the radiator is too small or becomes dirty or restricted, the engine will overheat, due to inadequate heat exchange. It should be noted that most vehicles equipped with automatic transmissions circulate their transmission fluid through a separate heat exchanger inside the radiator to help to warm the transmission up on cold days and to maintain consistent temperature after the vehicle has warmed up.
2.	Upper Radiator Hose	The radiator hoses are rubber hoses that provide a flexible connection between the engine and radiator.
3.	Lower Radiator Hose	The radiator hoses are rubber hoses that provide a flexible connection between the engine and radiator.
4.	Radiator Cap	The radiator cap is responsible for maintaining the cooling system's pressure. Late-model vehicles have closed cooling systems, which means that they use a radiator cap to control the coolant level along with a recovery bottle. The radiator cap has a seal that comes in contact with a surface inside the top of the radiator or coolant reservoir. There is a spring in the cap that holds pressure against this sealing surface. The pressure is usually between 13 and 16 psi. When the engine is cold, the cooling system has no pressure in it. As the engine warms up, the coolant expands. When the pressure in the system exceeds the caps rating, the excess coolant will be pushed out to a recovery bottle. This expansion will also push any air in the system out. In turn, when the engine is shut down, the coolant retracts as it cools. This creates a vacuum in the cooling system. The coolant is drawn back out of the recovery bottle until the system is full. There are really no moving parts involved, just a controlled use of expansion and contraction. If the system is otherwise leak-free, it will be full at all times and free of air, which causes accelerated deposits that can restrict the tubes of the radiator. One other advantage of running a pressurized cooling system is that coolant under pressure has a higher boiling point.
5.	Water Pump	If you can imagine an old paddle-wheel boat, you have a pretty good idea what the inside of a water pump looks like. One of the radiator hoses is connected to the inlet side of the water pump. The pump pulls coolant from the cooled side of the radiator and pushes it out through passages that travel through internal passages in the cylinder head and block. These passages are cast around all of the hot parts of the combustion process like the cylinders and the combustion chambers in the heads. The shaft that connects to the "paddle wheel," called an impeller, goes to the outside of the engine and has a pulley attached to it that is driven by a belt. This shaft has seals and bearings to allow it to spin at high speed while keeping the coolant in the engine. This is the location where most water pumps fail. Water pumps come in every shape and size, but all perform the same task.

COOLING SYSTEM EXPLODED VIEW—*Continued*

Diagram Number	Component Name	What It Does
6.	Thermostat	Probably the most misunderstood component of the cooling system, the thermostat does not keep the engine cool; it warms it up. The radiator is responsible for keeping the engine cool. To help the engine wear evenly and produce the least amount of tail pipe emissions, it must be kept at a constant and even temperature throughout. Engines make more excess heat when they are under load than they do when they are cruising under light loads. The radiator is a static device that is more efficient when there is more air flowing across it. Since ambient temperatures and the amount of air moving across the radiator can vary, something has to control it. This is the job of the thermostat. For most people, the thermostat is a device with a control knob that adjusts the temperature of their home. Here we need a different visual of the device. The thermostat for an engine is located inside the engine, submerged in coolant. It controls the amount of coolant that enters the radiator from the engine. When the engine is cold, a big spring inside the thermostat holds it closed to keep all of the coolant flowing in the engine. As the engine reaches operating temperature (usually about 195 degrees), the spring in the thermostat relaxes, allowing coolant to flow to the radiator for heat exchange. If the temperature drops, the thermostat closes the opening down. This keeps the temperature consistent.
7.	Thermostat Housing	Most engines have a thermostat housing which is either part of the engine or a separate part. They are often made of light-duty aluminum and have a radiator hose connection. They are subject to leaks and warpage in many applications.
8.	Cooling Fan	Since vehicles operate under a variety of conditions, there is no way to guarantee adequate airflow across the radiator when sitting in traffic or moving at slow speeds. The cooling fan is used to provide the wind when there is none. Cooling fans may be the old static fans that moved at the same speed as the water pump they were bolted to, They may be electric and controlled by a temperature switch or engine computer, or they may be mounted on a fan clutch.
9.	Condenser Fan	In electric fan applications, a motor is used to spin the fan when air conditioner is on.
Not Pictured	Fan Clutch	Fan clutches are used in applications where the fan is mounted to a moving part of the engine, usually the water pump. Fan clutches can be thermostatic or speed sensitive. In either case, they vary the amount of air the fan moves by slowing down the fan blade relative to engine speed. This allows the use of a large, highly effective blade for slow speeds that can be slowed down to improve engine performance when it is not needed.
Not Numbered	Heater Core	The heater core is a small radiator. Coolant from the engine is circulated through, and a blower blows across the core to provide hot air inside the vehicle.
Not Numbered	Overflow/Recovery Bottle	Coolant recovery/overflow bottles are the reservoirs that coolant moves into and out of as it expands and contracts. In many late-model vehicles, the bottle has been replaced by a totally closed system in which the bottle functions as the radiator pressure and off-gassing tank to help trapped air get out of the coolant.

COOLING SYSTEM EXPLODED VIEW—*Continued*

Diagram Number	Component Name	What It Does
Not Pictured	Antifreeze	The coolant in the engine comes in many types. It can be green, red, yellow, orange, or pink. The name *antifreeze* is only half the story. In addition to dropping the freezing point of water to around –40°F it also helps to keep from boiling clear up to 240°F, or more, depending on system pressure. The coolant is mixed with water in a 50/50 mixture in almost all applications. The coolant has the additional responsibility of lubricating the water pump, protecting the metal components of the engine, and improving the temperature transfer capabilities of the components it comes in contact with.

Task C.2.4

Provide basic use and installation instructions.

Helping DIY customers verify the part they think they need can save you lots of time in the long run. Here are some examples: To check the condition with the engine not running, grasp the fan blades, or the water pump hub, and try to move the shaft from side to side to check for looseness in the water pump bearing. If any side-to-side movement is noted in the bearing, water pump replacement is required.

Check for coolant leaks, rust, or residue at the water pump drain hole in the bottom of the pump and at the inlet hose connected to the pump. When coolant is dripping from the pump drain hole, replace the pump. The water pump may be tested with a pressure tester connected to the radiator filler neck.

A thermostat that is stuck open reduces coolant temperature and does not cause high coolant level in the recovery tank. The radiator cap should be inspected for a damaged sealing gasket or vacuum valve. If the pressure cap is damaged, the engine will overheat and coolant will be lost to the coolant recovery system. Under this condition, the coolant recovery tank becomes overfilled. If the cap valve is sticking, a vacuum may occur in the cooling system after the engine is shut off and the coolant temperature decreases. This vacuum may cause collapsed cooling system hoses. A pressure tester may be used to test the pressure cap and the entire cooling system. The coolant level should be at the appropriate mark on the recovery container, depending on engine temperature.

3. Fuel Systems (3 Questions)

Tasks C.3.1 C.3.2 C.3.3

Identify major and related components and their function.

The fuel system is the system that delivers fuel to the engine cylinders. It consists of a fuel tank and lines, gauge, fuel pump, carburetor or injectors, and intake manifold or fuel rail.

The fuel system supplies a combustible mixture of gasoline and air to the engine's cylinders. To do this, the fuel system must store the fuel, deliver fuel to the metering devices, atomize and mix the fuel with air, and vary the proportion of fuel-to-air to satisfy the many load requirements of the engine.

For Tasks C.3.1 through C.3.3 we are providing you with an itemized list of the major components in a format familiar to you as a parts specialist. Each item in the diagram has a matching description in the table below it.

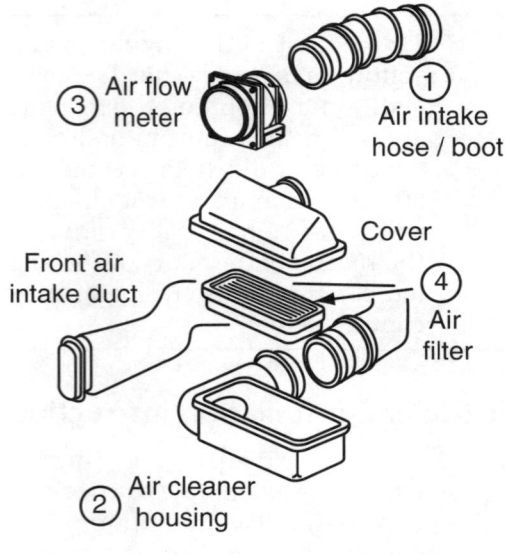

Induction System

INDUCTION SYSTEM

Diagram Number	Component Name	What It Does
1.	Air Intake Hose/Boot	This hose connects the air cleaner to the throttle body in most fuel-injected vehicles. It may have connections to the PCV system. Because they provide the flex between the body and the engine, breakage occurs, causing strange drivability problems.
2.	Air Cleaner Housing	Holds the air filter. There are as many different configurations as there are vehicles, but they all hold the air filter. Some manufacturers mount the air meter or mass airflow sensor on the air cleaner, as in this picture.
3.	Air Flow Meter	There are two types of meters commonly in use, the Vane type and the Mass air flow meter. The vane type meter has a moving flap in the incoming air tract that opens and closes, based on how much air the engine is using. The computer uses this signal to calculate how long to open each injector. The other type of meter uses a heated wire across the intake stream. When air crosses this wire or grid, the wire cools off. This change in temperature is used by the computer to calculate how much air the engine is using and how much fuel to deliver.
4.	Air Filter	The air filter element has the job of removing dirt and other contaminants from the incoming air to the engine. All the various designs perform this same function.

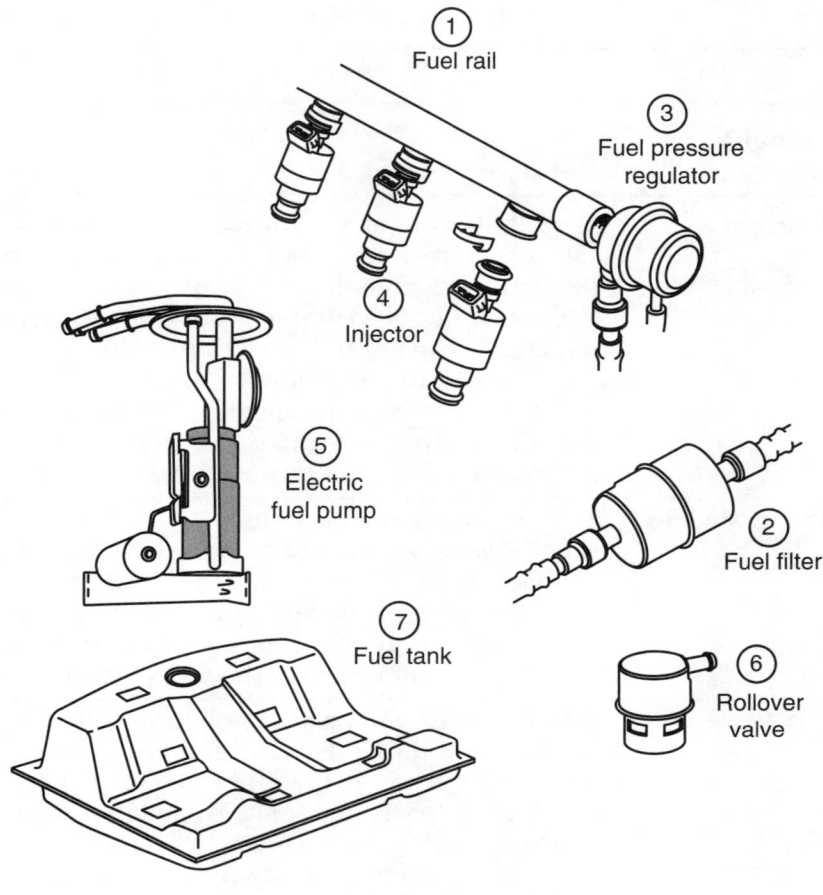

Fuel System

FUEL SYSTEM

Diagram Number	Component Name	What It Does
1.	Fuel Rail	The fuel rail connects the fuel lines to the fuel injectors. In many applications, the fuel pressure regulator is attached to the fuel rail. The fuel injectors are retained in the fuel rail by O-rings and, in some applications, clips. This is traditionally not a part that wears out.
2.	Fuel Filter	Fuel filters contain the filtering media used to remove dirt and debris from the fuel system after the fuel pump, but before the fuel injectors.
3.	Fuel Pressure Regulator	Most of the time the fuel pressure regulator is located in the engine compartment. It resides in the return side of the fuel line and keeps the fuel pressure at the necessary level by restricting the amount of fuel returned to the tank. The control for the pressure may be a fixed spring inside the regulator or vacuum from the engine, or it may be controlled by the PCM in the returnless system. These systems do not have a return line, but rather vary the speed of the electric fuel pump. In this situation, the regulator is a monitoring device telling the computer how much pressure is being made.

FUEL SYSTEM—*Continued*

FUEL SYSTEM

Diagram Number	Component Name	What It Does
4.	Fuel Injector	A fuel injector is an electrical component. It is a high-speed solenoid. When the computer completes the electrical circuit to it, the Pintle, which is a small needle valve, is lifted up off the seat, and fuel sprays out of the tip in a nice cone shape. When the computer releases the circuit, the valve closes. This is measured by the technician in milliseconds. An injector is typically open for 1–2 milliseconds at idle and 12–15 milliseconds under heavy load. This rapid movement and the cycle from cold to hot and back again are the most common causes of injector failure.
5.	Fuel Pump Module	The fuel pump module houses the fuel pump and, in many cases, the fuel gauge sending unit, along with any plumbing necessary to complete the fuel return to the tank. Most manufacturers sell this as an assembly. Some fuel pumps are sold separately and are installed in the original module or mounted in other places on the vehicle. The in-tank unit is, by far, the most common design.
6.	Rollover Valve	It is likely that you have not run into this component. This device serves two purposes. First, it keeps fuel from running out of the fuel tank vent line in the event of a vehicle roll over. Second, it is as a check valve for the evaporative emission system. Evaporative emissions systems in fuel-injected vehicles are really designed to control fuel vapors that might leave the tank as the engine warms the fuel from the constant circulation through the fuel rail and back to the tank. In vehicles built from 1996 on, the evaporative system is controlled and monitored by the engine management computer or PCM (Powertrain control module). The fuel tank is a critical and major component of this system. The largest evaporative system line is usually connected to the roll-over valve and then to the engine on the other end. A solenoid-controlled vacuum valve is used by the PCM to apply vacuum to the tank for testing and to draw vapor into the engine to be burned.
7.	Fuel Tank	In late model vehicles, the fuel tank is more than just a holding compartment for fuel. It functions as a surge tank to manage the changes in fuel temperature. As we mentioned before, it usually contains the fuel pump and a sending unit. It also has baffles inside to help the fuel stay as close to the fuel pump pick up as possible.
Not Pictured	PCV Valve	If the positive crankcase ventilation (PCV) valve is stuck in the open position, excessive airflow through the valve causes a lean air/fuel ratio (in carbureted vehicles) and possible rough idle operation or engine stalling. When the PCV valve or hose is restricted, excessive crankcase pressure forces blowby gases through the clean air hose and filter into the air cleaner. Worn rings or cylinders cause excessive blowby gases and increased crankcase pressure, which force blowby gases through the clean air hose and filter into the air cleaner. A restricted PCV valve or hose may result in the accumulation of moisture and sludge in the engine and engine oil.

FUEL SYSTEM—*Continued*

FUEL SYSTEM

Diagram Number	Component Name	What It Does
Not Pictured	Carburetor	The carburetor made its last appearance in vehicles around 1991. There are still plenty of vehicles running that use them, so it is important to know how they work. Carburetors use the engine's vacuum to draw fuel into mixing wells, where the fuel is atomized and delivered through discharge venturis inside the throttle bores where it mixes with air and enters the engine through the intake manifold. Many of the external components of carburetors, including choke components, choke pull-offs, and fuel filters are common service items.

Task C.3.4   **Provide basic use and installation instructions.**

The components of the fuel system are frequently blamed for problems related to other parts of the engine or ignition system. To avoid selling a part that will ultimately be returned, you can use the information you gain in these charts and through your experience to help to direct customers to the correct decision, whether that is adding a book to the sale or directing them to one of your professional customers if you determine they are over their ability threshold.

4. Ignition Systems (3 Questions)

Tasks C.4.1 C.4.2 C.4.3   **Identify major and related components and their function.**

The ignition system is an electrical system. It is made up of two parts. The part most obvious to customers is the ignition switch. From a simple switch to turn the power to the engine on, the ignition switch has grown to being master control switch for the entire vehicle. The second part of the ignition system is what has traditionally been called the ignition system. This ignition system includes the components that are associated with tune-up parts. With vehicle maintenance seeing so many changes in the last decade, the tune-up has become a very gray area that does not really fit anymore. In the next area, we are going to look at the components of both a conventional distributor electronic ignition and distributorless ignition system.

For Tasks C.4.1 through C.4.3, we are providing you with an itemized list of the major components in a format familiar to you as a parts specialist. Each item in the diagram has a matching description in the table that follows.

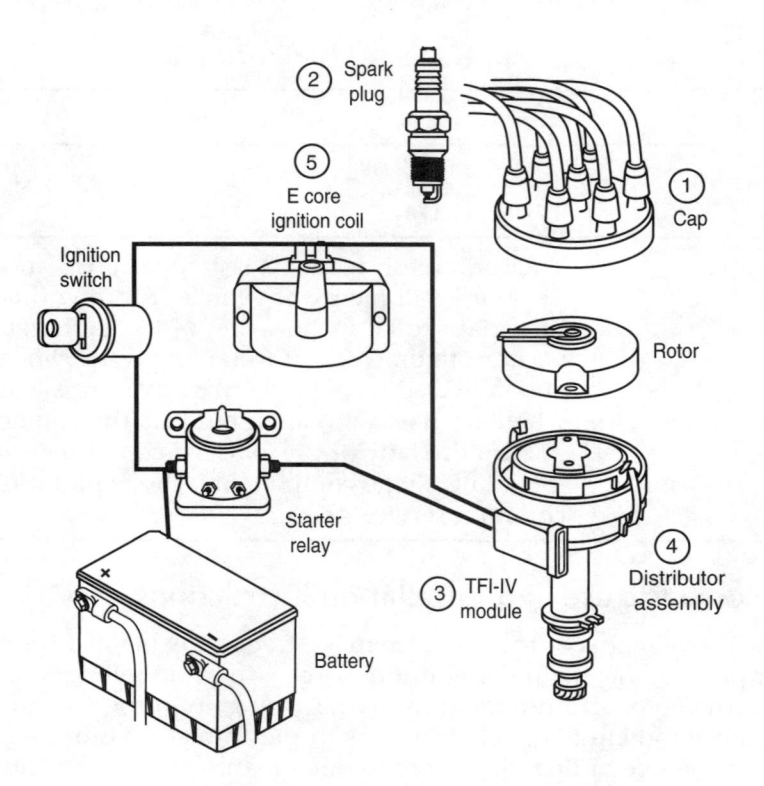

Electronic Ignition with Distributor

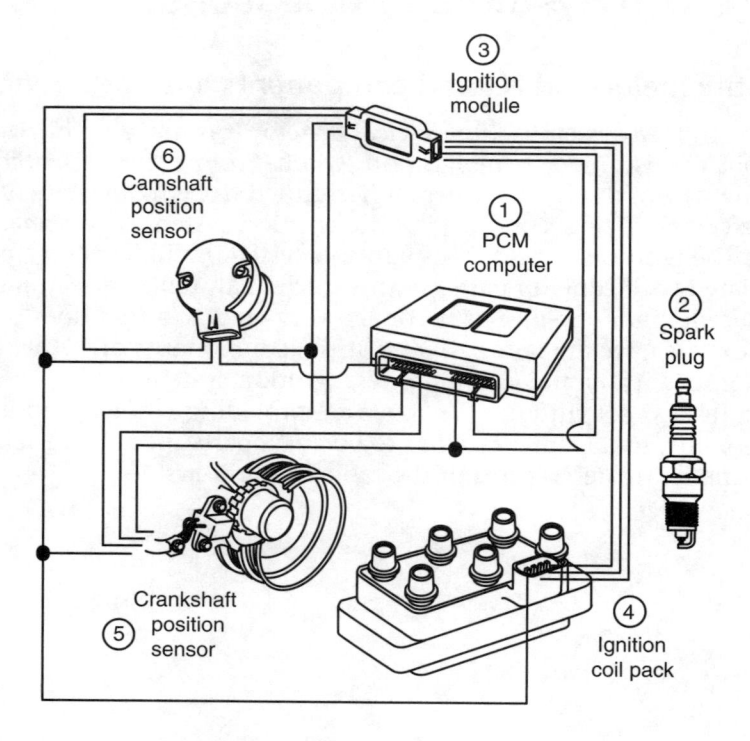

1. Cap
2. Spark plug
3. TFI-IV module
4. Distributor assembly
5. E core ignition coil

Typical Distributorless Ignition

1. PCM computer
2. Spark plug
3. Ignition module
4. Ignition coil pack
5. Crankshaft position sensor
6. Camshaft position sensor

ELECTRONIC IGNITION SYSTEM WITH DISTRIBUTOR

Diagram Number	Component Name	What It Does
1.	Distributor Cap and Rotor	Probably the most familiar component, with the exception of the spark plug, The distributor cap has a small button in the center that carry current from ignition coil to the center electrode of the rotor. The rotor is attached to a shaft in the distributor that turns at the same speed as the engine's valve train. When the tip of the rotor passes under one of the contacts in the distributor cap, the electricity it is carrying finds a ground through the spark plug at the other end of an ignition wire that is attached to the plug and the cap. Due to the very high voltage levels that jump across these contact points, the cap and rotor are subject to deterioration over time.
2.	Spark Plug	Spark plugs are very simple devices. They provide a ground for the ignition system through the threaded body that bolts into the cylinder head. When current is applied to them, a small lightning bolt jumps across the inner electrode that is insulated from the outer shell to the outer electrode attached to the spark plug shell. This spark starts a chemical reaction in the combustion chamber that causes the fuel and air mixture to oxidize very rapidly, heating it and causing expansion. Spark plugs may have electrodes made of or coated with copper, platinum, or titanium to help them last longer. When spark plugs wear, the electrodes that are very square to the eye start to round and erode. This wear will cause the ignition system to work harder to make that spark jump the gap.
3.	Fuel Pressure Regulator	The fuel pressure regulator is located in the engine compartment most of the time. It resides in the return side of the fuel line and keeps the fuel pressure at the necessary level by restricting the amount of fuel returned to the tank. The control for the pressure may be a fixed spring inside the regulator or a vacuum from the engine, or it may be controlled by the PCM in the returnless system. These systems do not have a return line but rather vary the speed of the electric fuel pump. In this situation the regulator is a monitoring device telling the computer how much pressure is being made.
4.	Distributor	The distributor has a mechanical connection to the engine and turns at the speed of the valve train. In earlier vehicles that had a mechanical spark advance, the distributor had a vacuum-controlled advance mechanism, as well as centrifugal advance. Modern vehicles that are equipped with distributors use computer-controlled timing and were basically stop-gaps to the current distributorless systems.
5.	Ignition Coil	The ignition coil takes a low voltage pulsing signal from the ignition module and amplifies is from about 9–12 volts to 5,000–40,000 volts and sends it to the spark plugs.

ELECTRONIC DISTRIBUTORLESS IGNITION SYSTEM

Diagram Number	Component Name	What It Does
1.	PCM (computer)	The PCM is responsible for managing spark timing in many distributor and all distributorless systems. In the DIS system the crank and sometimes the camshaft position sensors are used to synchronize spark plug firing to fuel injection and crankshaft events. The PCM may use many inputs like the coolant temperature sensor, a knock sensor, throttle position, mass air flow, EGR system, etc., to calculate the modifications to the timing program built into its memory.
2.	Spark Plug	Spark plugs are very simple devices. They provide a ground for the ignition system through the threaded body that bolts into the cylinder head. When current is applied to them, a small lightning bolt jumps across the inner electrode that is insulated from the outer shell, to the outer electrode attached to the spark plug shell. This spark starts a chemical reaction in the combustion chamber that causes the fuel and air mixture to oxidize very rapidly heating it and causing expansion. Spark plugs may have electrodes made of or coated with copper, platinum, or titanium to help them last longer. When spark plugs wear, the electrodes that are very square to the eye start to round and erode. This wear will cause the ignition system to work harder to make that spark jump the gap.
3.	Ignition Module	The ignition module receives a low-voltage-pulsed signal from either the distributor pick-up coil, the PCM, or the crankshaft position sensor and amplifies it, anywhere from 5 volts or less to a 12-volt signal that is sent to the coil(s), which amplifies it again to several thousand volts to fire the spark plug. In computer-controlled applications, the ignition module usually acts as the middle management between the PCM and the ignition coil, taking commands from the PCM for timing and often providing the fixed or "base" timing necessary to start the engine and keep it running until the computer takes over the responsibility.
4.	Ignition Coil Pack	The ignition coil takes a low voltage pulsing signal from the ignition module and amplifies it from about 9–12 volts to 5,000–40,000 volts and sends it to the spark plugs. In the distributorless ignition system, there are multiple coils. In some applications, cylinders share a coil, while coil on plug systems have a coil for each spark plug that connects directly to the spark plug.
5.	Crank Position Sensor	In the distributorless system, the PCM or computer must have a way to determine what position the crank is in rotation so that it knows when to fire the spark plugs. This is accomplished by using a degreed wheel on the crank or flywheel with a special signature that tells the PCM which location is the number-one cylinder. This helps the PCM synchronize the spark plugs to the crankshaft position. Because the crank turns two revolutions for each revolution of the valve train, spark plugs are fired twice during a complete cycle: once during the power stroke and once during the exhaust stroke.

ELECTRONIC DISTRIBUTORLESS IGNITION SYSTEM—*Continued*

Diagram Number	Component Name	What It Does
6.	Cam Position Sensor	In some systems, an additional sensor, called the cam position sensor, takes the place of the special signature on the crank sensor wheel. This is used to determine the exact position of the camshaft and number-one cylinder. This added sensor is most commonly found in systems that use sequential multi-port fuel injection when the PCM needs to fire both the injector and the spark plugs at precise times in reference to crank position.

Task C.4.4 **Provide basic use and installation instructions.**

Probably the most valuable information a parts specialist can provide to a customer when working with the ignition system is the value of using the correct parts and installation practices. While there is no way for you to know every vehicle's idiosynchracies, you can help to steer the customer in the right direction by recommending additional necessary items when replacing items like spark plugs—things like a gapping tool or anti-seize for the threads. The gap specification and the correct type of spark plug are simple items that make the difference.

5. Exhaust Systems (2 Questions)

Tasks C.5.1 C.5.2 C.5.3 **Identify major and related components and their function.**

The exhaust system is a major system in a vehicle. The exhaust system is designed to channel toxic exhaust fumes away from the passenger compartment, to quiet the sound of the exhaust pulses, and to burn, or catalyze, emissions in the exhaust. On some engines, the exhaust system performs several other functions. Some vehicles have a stove on the exhaust manifold that is used to heat the incoming air. A hot-air pipe or hose leads up from the manifold to the air cleaner snorkel and carries heated air to mix with the normal intake air.

Some engines use manifold heat to warm a section of the intake manifold. An exhaust crossover passage directs exhaust under the intake manifold, warming the air/fuel mixture.

For Tasks C.5.1 through C.5.3, we are providing you with an itemized list of the major components in a format familiar to you as a parts specialist. Each item in the diagram on the next page has a matching description in the table that follows.

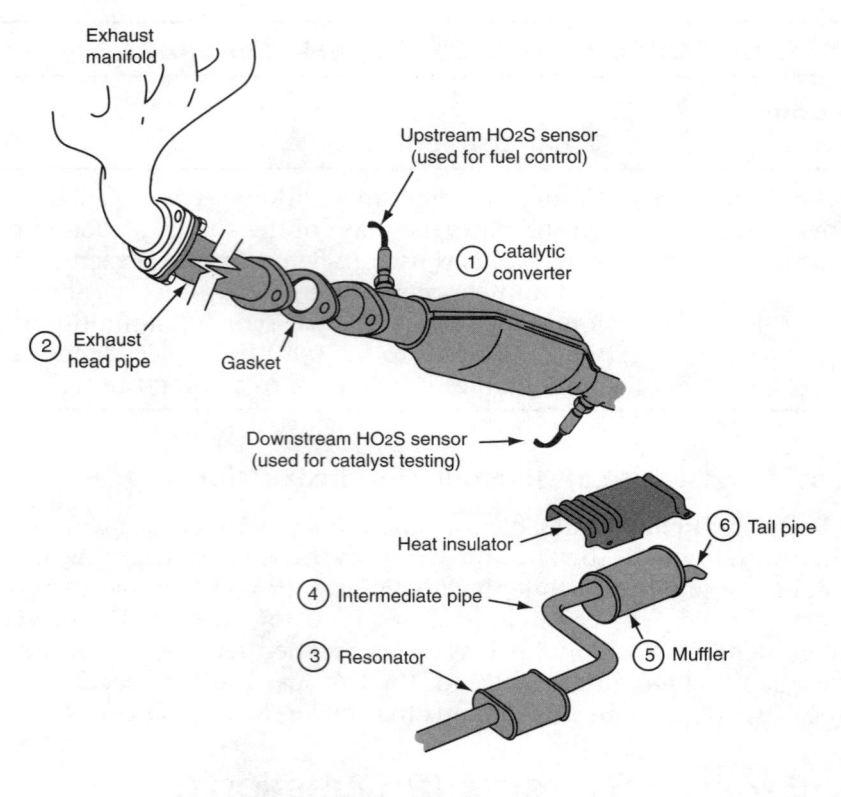

Exhaust Systems

EXHAUST SYSTEMS

Diagram Number	Component Name	What It Does
1.	Catalytic Converter	The most complex component of the exhaust system is an emission component with no moving parts. The catalytic converter is located at the front of the vehicle, usually as far forward as possible. The catalytic converter has the job of removing or reducing several byproducts of combustion. These components are carbon monoxide (an odorless, invisible, and deadly gas), hydrocarbons (small quantities of unburned fuel) and NOX or oxides of nitrogen (not to be mistaken for nitrous oxide), which are responsible for photochemical pollution (acid rain). These emissions are removed by passing the exhaust stream over a bed of precious metals that cause a chemical reaction with the exhaust when they are hot. The outcome in a perfect world is basically water. This is why you often see large amounts of water running out of the tailpipe of late-model vehicles.
2.	Head Pipe	The head pipe is the pipe directly behind the exhaust manifolds on the engine. It often has the catalytic converter built in.
3.	Resonator	Some exhaust systems have a resonator in them to help remove some of the engine's noise. These may be located before or after the muffler, depending on what the engineers were trying to accomplish. You can think of it as a pre- or post-muffler muffler.

EXHAUST SYSTEMS—Continued

Diagram Number	Component Name	What It Does
4.	Intermediate Pipe	The pipes between head pipe and the muffler are known as intermediate pipes. On V-engines with dual exhaust, sometimes there is a pipe that joins the two sides of the engine in the intermediate pipe. This pipe is called a crossover or H-pipe.
5.	Muffler	The muffler is a resonating chamber that cancels out loud noises from the engine. There is usually a perforated tube that runs through the muffler. The perforations allow sound to be reflected to the chambers within the muffler. These are sometimes filled with damping material like steel wool or ceramic panels.
6.	Tail Pipe	Just like it sounds, this is the pipe that exits the exhaust out from under the vehicle at the rear or side of the vehicle.

Task C.5.4

Provide basic use and installation instructions.

Many original mufflers and catalytic converters are integral with the interconnecting pipes. When these components are replaced, they must be cut from the exhaust system with a cutting tool. The inlet and outlet pipes on the replacement muffler or converter must have a 1.5 inch (38 mm) overlap on the connecting pipes. Before cutting the pipes to remove the muffler or converter, measure the length of the new component and always cut these pipes to provide the required overlap.

Normal exhaust noise contains steady pulses at the tailpipe. A puff noise in the exhaust at regular intervals usually indicates a cylinder misfire caused by a compression, ignition, or fuel system defect. Erratic exhaust pulses at the tailpipe indicate a rough idle condition caused by ignition or fuel system defects. Excessive exhaust noise indicates a leak in the exhaust system.

A high-pitched squealing noise during hard acceleration may be caused by a small leak in the exhaust system, particularly in the exhaust manifolds or exhaust pipe.

6. Emissions Control Systems (4 Questions)

Tasks C.6.1 C.6.2 C.6.3

Identify major and related components and their function.

Emission control systems encompass nearly every engine control system. We have discussed the evaporative system in the fuel system section and the catalytic converter in the exhaust. Let's round out your knowledge of the basics of emission control systems by looking at a general computer controlled system and some of its possible emission control devices.

Because different vehicles require different types of equipment, you will not find specific system questions here, but you will greatly benefit from understanding how the systems work. These can be very expensive components to repair, and they seldom have much effect on the way the vehicle runs. This is why the part specialist must be able to help the customer understand the implications of driving a vehicle with the Malfunction Indicator Light (MIL) on.

The modern engine management system is an amazingly adaptable and dependable system. It makes vehicles start and run much better and more consistently than pre-computer controlled vehicles. It might surprise you to know that the original reason to have on-board computers was actually to control emissions. The easiest way to understand these systems is to relate them to our own bodies. The computer is like our brain. It has a vast store of information and "programming" in it. It can adapt to changes in weather, driving style, road conditions, and even vehicle wear. Unlike the humans who designed them, Powertrain Control Modules (PCM) do not have to make many mistakes to learn, and they can learn in milliseconds. That doesn't mean that when they get poor inputs they don't make bad decisions.

So, how do these little boxes come to all these rapid and accurate decisions? One easy way to think of it is to imagine that they use electronics to replicate human senses. Let's start with smell: the computer uses one or more oxygen sensors mounted in the exhaust system to sniff the exhaust stream to see if it is delivering too much or too little fuel. It can even use this information to confirm a misfire condition or a bad catalytic converter. In reality, its sense of smell is just an electrical signal sent to its Oxygen Sensor inputs. Moving on to feel or touch, the computer uses several sensors to this end. A coolant and air temperature sensors tell it the temperature. The mass airflow sensor feels the amount of air the engine is breathing and tells the PCM. A crank and cam sensor help the PCM see the engine's speed and track misfires. Many engines use a set of electronic ears to listen for pinging or detonation. The PCM uses this information to adjust timing and protect the engine from damage. The PCM "sees" the position of the driver's foot on the throttle by tracking the throttle position sensor. There are many other inputs to the PCM. They are included in the task list.

The PCM takes all of these inputs and responds to provide the best emissions and drivability to the driver. This includes varying timing and fuel delivery many times per second, tracking inputs like air conditioner commands and power steering inputs to adjust idle speeds, and responses to load to vary the application of emission control devices like air pumps, evaporative systems, and EGR valves.

The EGR or Exhaust Gas Recirculation System is used to reintroduce a controlled amount of exhaust back into the combustion process. Because exhaust is basically an inert gas that will not create any heat during combustion, it helps to cool the process down under light loads. This serves two purposes. First, it lowers NOX emissions and, second, it helps to cool down the internal parts of the top end of the engine like cylinder heads, pistons, and valves, which can help to eliminate pinging that can occur under these leaner fuel conditions. Pinging or detonation occurs when the air and fuel mixture fires before the piston is all the way up on the compression stroke. Since the piston is connected to the crank, it must keep going up, but the force created by the false ignition process makes a knocking or pinging sound as the flame front hits the top of the piston like a hammer. When this happens, the engine is misfiring and creating large amounts of all emissions.

Air pump systems have been in use since 1969 when some of the manufacturers used them to get their high performance engines past the California emissions standards. This system pumps ambient air into the exhaust stream to help finish combustion of any burnable byproducts in the exhaust system. Air injection is necessary for some types of catalytic converters to be efficient. The system is made up of a belt drive or electric pump on the engine, lots of plumbing, and some one-way check valves to keep exhaust from getting into the pump. These systems are controlled by systems ranging from simple vacuum valves to complex strategies built into the PCM. These systems allowed manufacturers to continue to use older engine designs that were not as efficient and maintain lower emission standards. Most vehicles no longer use air pumps, as the newer engine designs are cleaner.

Task C.6.4

Provide basic use and installation instructions.

The information in C.6.1 through C.6.3 should give you all the information you need to assist customers with engine emission and management systems.

7. Manual Transmission/Transaxle (2 Questions)

Tasks
C.7.1
C.7.2
C.7.3

Identify major and related components and their function.

The manual transmission uses a clutch and manual shifting to change gears. The gear train is simplified in the diagrams. We will look at a basic transaxle to identify the components. Most of the components in the transaxle are the same in a rear wheel drive transmission, except that the differential components are housed inside the transaxle and are external on a rear or 4WD application.

For Tasks C.7.1 through C.7.3 we are providing you with an itemized list of the major components in a format familiar to you as a parts specialist. Each item in the diagram has a matching description in the table below it.

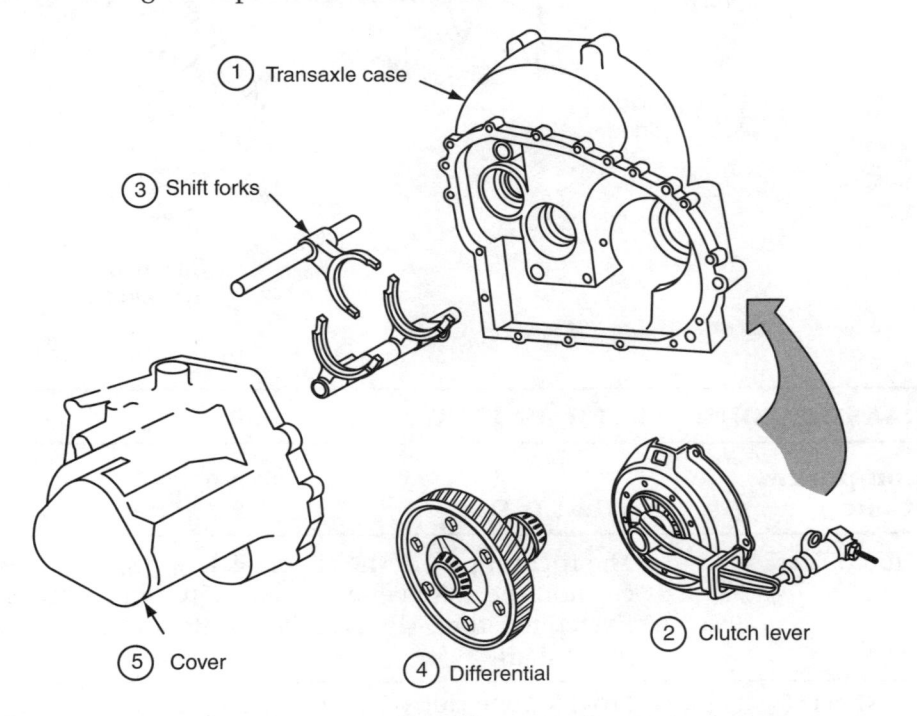

Manual Transaxle (many components removed for clarity)

MANUAL TRANSMISSION/TRANSAXLE

Diagram Number	Component Name	What It Does
1.	Transaxle Case	All of the components attach or are housed by this assembly.
2.	Clutch Lever	The external connection to either the hydraulic cylinder or cable that operates the clutch inside the bell housing.
3.	Shift Forks	Manual transmissions use forks that straddle the synchro/gear assemblies to move the synchro from gear to gear causing a shift
4.	Differential	In a front wheel drive, the differential is usually inside the case and driven by a final drive gear from the transmission gear train that engages the large ring gear.
5.	Cover	An access point is available on some transaxles. The final drive gears are often in this cover.

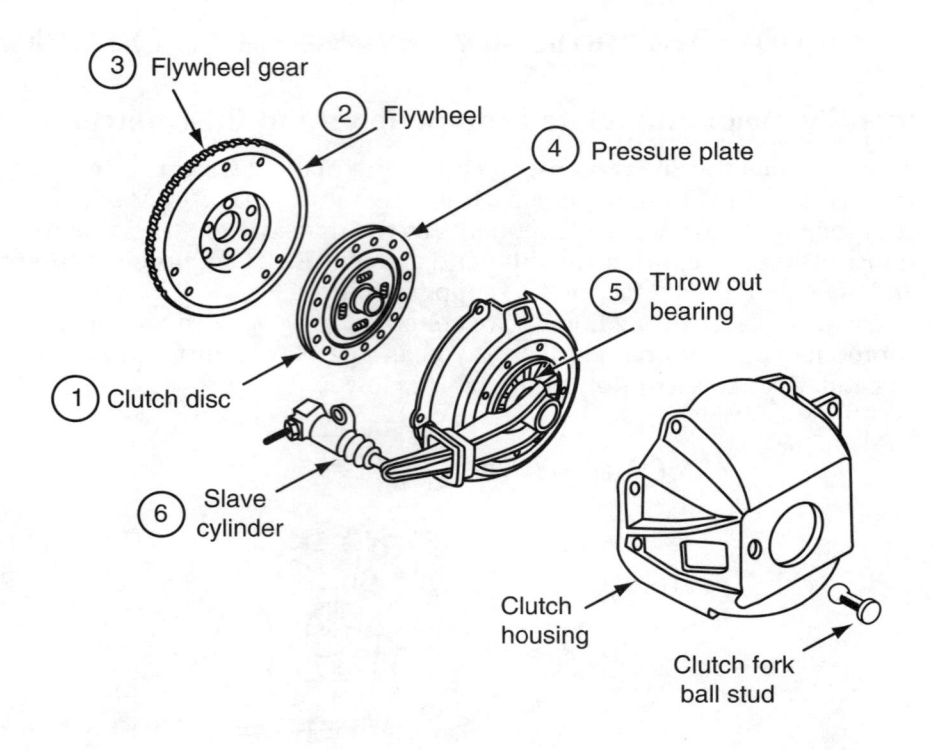

MANUAL TRANSMISSIONS: CLUTCH ASSEMBLY

Diagram Number	Component Name	What It Does
1.	Clutch Disc	The friction part of the clutch. The springs in the center are there to cushion the clutch engagement. The clutch disc is two-sided and riveted together. The disc slides onto and drives the transmission input shaft.
2.	Flywheel	Provides one side of the disc-clamping surface. The flywheel is machined smooth before assembly and bolts to the rear of the crankshaft.
3.	Flywheel Gear	This gear is pressed onto the flywheel and is the gear the starter engages when turning the engine over during starting.
4.	Pressure Plate	Provides the clamping force and is actuated by the clutch fork to release the clutch.
5.	Throw Out Bearing	This bearing is a sealed bearing that rides against the pressure plate actuating fingers during disengagement. Many late-model vehicles apply light pressure so that the throw out bearing is in constant contact with the pressure plate.
6.	Slave Cylinder	Some clutches use a cable to connect the pedal assembly to the clutch actuation fork. Others use a hydraulic system similar to brakes. The slave cylinder may be inside or outside the bell housing. The slave cylinder expands to push on the clutch fork when the pedal is depressed and the clutch master cylinder generates pressure.

Task C.7.4 **Provide basic use and installation instructions.**

Worn input shaft splines may cause the clutch disc to stick on these splines, resulting in improper clutch release. The gears on the input shaft should be inspected for cracks and pitted, worn, or broken teeth. The bore in each gear and the matching surface on the input shaft should be inspected for roughness, pits, and scoring. Inspect the needle bearings mounted between the gears and the input shaft for roughness and looseness. Inspect the output shaft bearings in the transaxle case for roughness or looseness.

8. Automatic Transmission/Transaxle (2 Questions)

Tasks
C.8.1
C.8.2
C.8.3

Identify major and related components and their function.

The modern automatic transmission is a fairly complex unit. Let's boil its operation down to the basics. Just like the manual transmission, it provides different gearing to help the engine move the vehicle. Unlike the manual, it uses a series of bands that grab hold of spinning drums and a planetary gear that can create several different gear ratios, depending on how it is manipulated by the transmission. The automatic transmission is controlled hydraulically and, instead of a clutch that must be engaged and disengaged, it uses a hydraulic coupling device known as a torque converter that varies the amount of engine torque and engagement. The internal controls of the transmission are fed hydraulic pressure by pumps inside the transmission and controlled by the valve body that directs shifts by changing pressures and directing fluid through passages. Below is a diagram of some of the major components of the automatic transmission along with a chart of their names and purposes. You will see many similar components with our manual chart.

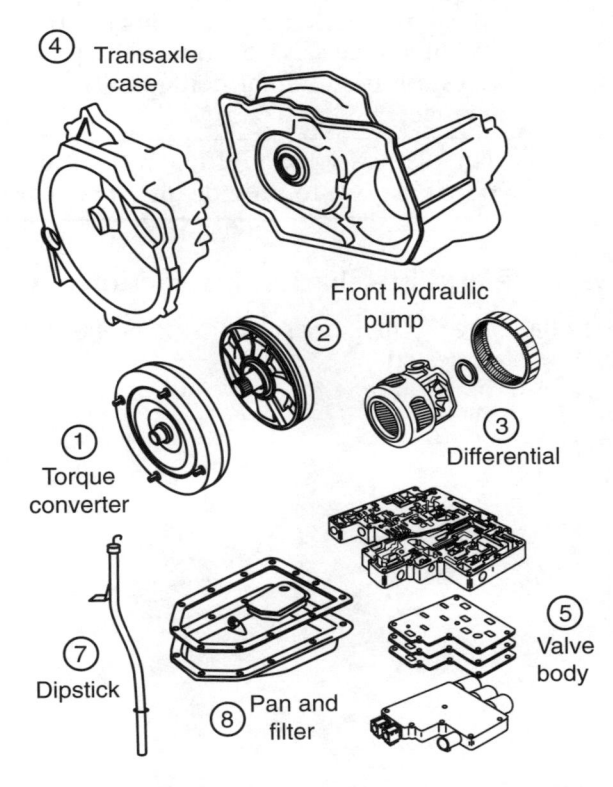

Automatic Transaxle Assembly (many components removed for clarity)

AUTOMATIC TRANSMISSION/TRANSAXLE

Diagram Number	Component Name	What It Does
1.	Torque Converter	A hydraulic coupling device that connects the engine to the transmission. Most late-model automatics have an electric locking clutch inside the converter that makes a direct connection between the engine and the transmission during light load cruising. This drops RPM and helps with gas mileage.
2.	Front Hydraulic Pump	This is one of the pumps in a transmission. It is the only pump in this transmission; some have a rear pump as well.
3.	Differential	In a front wheel drive, the differential is usually inside the case and driven by a final drive gear from the transmission gear train that engages the large ring gear.
4.	Transaxle Case	All of the components attach to or are housed by this assembly.
5.	Valve Body	The hydraulic control unit. Newer electronic transmissions have solenoids within the valve body that are run by the PCM.
Not pictured	Side Cover	Usually a cover for the valve body in front wheel drive applications. Most rear wheel drive vehicles have their valve body in the bottom of the transmission in the oil pan area.
7.	Oil Dipstick	Automatic transmissions have a fluid level dipstick, while most manual gearboxes do not.
8.	Pan and Filter	Most automatic transmissions use filters that can be replaced as maintenance service. Some vehicles are equipped with a screen that does not require maintenance. These vehicles simply require fluid changes.
Not pictured	Planetary	The planetary is a multiple ratio gear set used in automatic transmissions to provide different gear ratios in a compact package.

Task C.8.4 Provide basic use and installation instructions.

The biggest challenge to the parts specialist is properly identifying the transmission and the part that is requested.

Overview of the Task List | Automotive Parts Specialist (Test P2) 43

9. Drivetrain Components (Includes Driveshafts, Half Shafts, U-Joints, CV Joints, and Four-Wheel-Drive Systems) (2 Questions)

Tasks
C.9.1
C.9.2
C.9.3

Identify major and related components and their function.

Drivetrain components include the rest of the components that connect the engine and transmission to the wheels. The components include: drive shafts, CV axles, differentials, and 4-wheel drive or all wheel drive transfer cases. There are so many designs we are going to take a look at generic versions of each to help you understand how they work and interrelate.

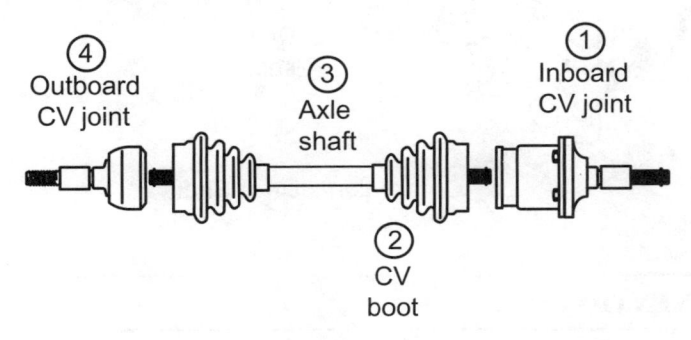

④ Outboard CV joint
③ Axle shaft
① Inboard CV joint
② CV boot

CV Drive Axle

DRIVE SHAFTS: CV AXLE

Diagram Number	Component Name	What It Does
1.	Inner CV Joint	The inner joint snaps into the differential side gears. This joint operates only in two planes. It must handle in and out movement, called plunge, as the vehicle moves up and down, changing the length of the driveshaft. It must also handle moving up and with the suspension.
2.	CV Boot	The major maintenance item on CV axles. The outer boots have the highest failure rate due to all the movement they must provide. The boot holds the lubricant in the joint and has bellows-shaped ridges that help it to move with changes in position.
3.	Axle Shaft	The axle shaft has splined areas at each end to engage and mount both CV joints. It may be hollow or solid steel.
4.	Outer CV Joint	The workhorse of the CV axle, this joint must operate in multiple planes at one time while rotating. The outer joint must turn with the wheels and move up and down with suspension. You will find CV axles used in front-wheel drive vehicles to both wheels, on front axles in late model 4-wheel drives with independent suspension, rear differential applications on mini 4-wheel drives, and even on some drive shaft applications.

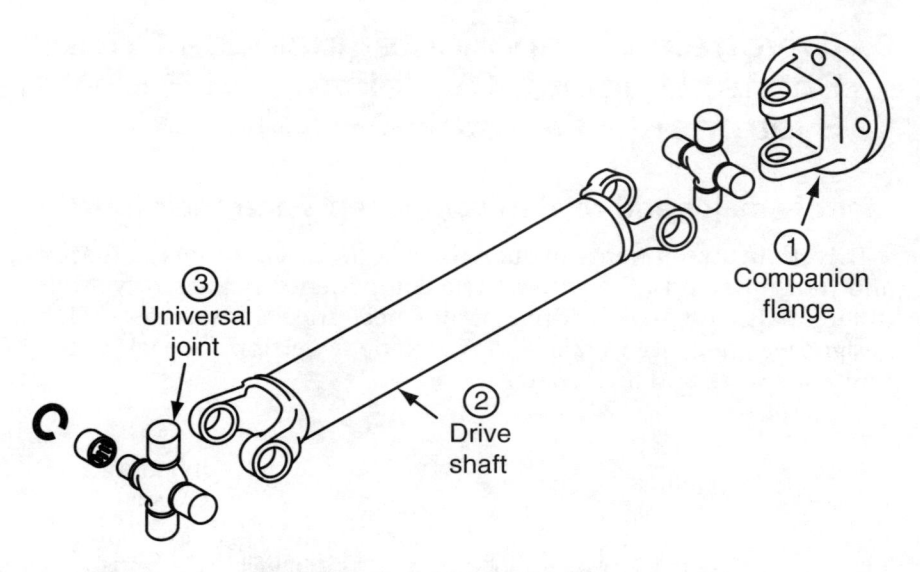

Conventional Drive Shaft

DRIVE SHAFT: CONVENTIONAL

Diagram Number	Component Name	What It Does
1.	Companion Flange	This type of flange bolts to a differential or transfer case flange. Many vehicles will place the actual u-joint bearing cups directly into a yoke on the differential or transfer case output and retain them with a u-bolt. This is becoming the standard approach because of the repeatability of installation, which helps to control vibration.
2.	Drive Shaft	This component provides the connection between the transmission or transfer case and the differential. They may be constructed of solid or hollow steel tubing, aluminum, or carbon fiber. The u-joints are pressed into the drive shaft and retained by clips in most applications.
3.	U-joints	U-joints allow the driveshaft to make a connection while rotating and moving up and down with the suspension. There are four cups that hold needle bearings that allow the u-joints to turn to accommodate the movement.

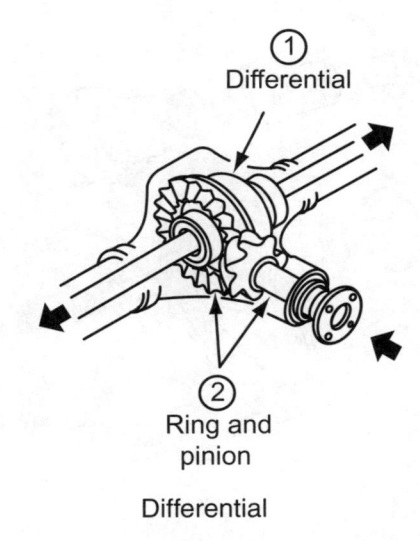

① Differential

② Ring and pinion

Differential

DIFFERENTIAL

Diagram Number	Component Name	What It Does
1.	Differential	Differentials may be found at either the front or rear of the vehicle. Front-wheel drive vehicles incorporate the differential into the transaxle. 4-wheel drives use a differential at each end of the vehicle and rear wheel drives have one at the rear of the vehicle. It is a safe conclusion that the name that precedes *drive* tells you where differentials are located. So what does the differential do? In our picture, you will see that the arrow points to a component. This is the carrier, and it contains the actual differential. Differential allows the drive wheels to turn at different speeds when the vehicle is turning a corner. In this situation, the outside wheel has a greater distance to travel than the inner wheel. If they were locked together, it would be a pretty jerky ride if the wheels could not travel at different speeds. Limited slip differentials use clutches to lock the two wheels together when going in a straight line; these slip to help to make turns.
2.	Ring and Pinion	In a conventional rear-wheel or 4-wheel drive vehicle, the driveshaft connects to one of two large gears, called the pinion gear. The gear that it meshes with changes the direction of the driveshaft rotation 90 degrees. It is called the ring gear. The ring gear is bolted to the carrier and causes the carrier to turn. In front-wheel drives, the engine's crank runs parallel to the drive axles. There is a form of final drive from the transmission part of the transaxle that drives the ring gear and carrier assembly.

Transfer Case

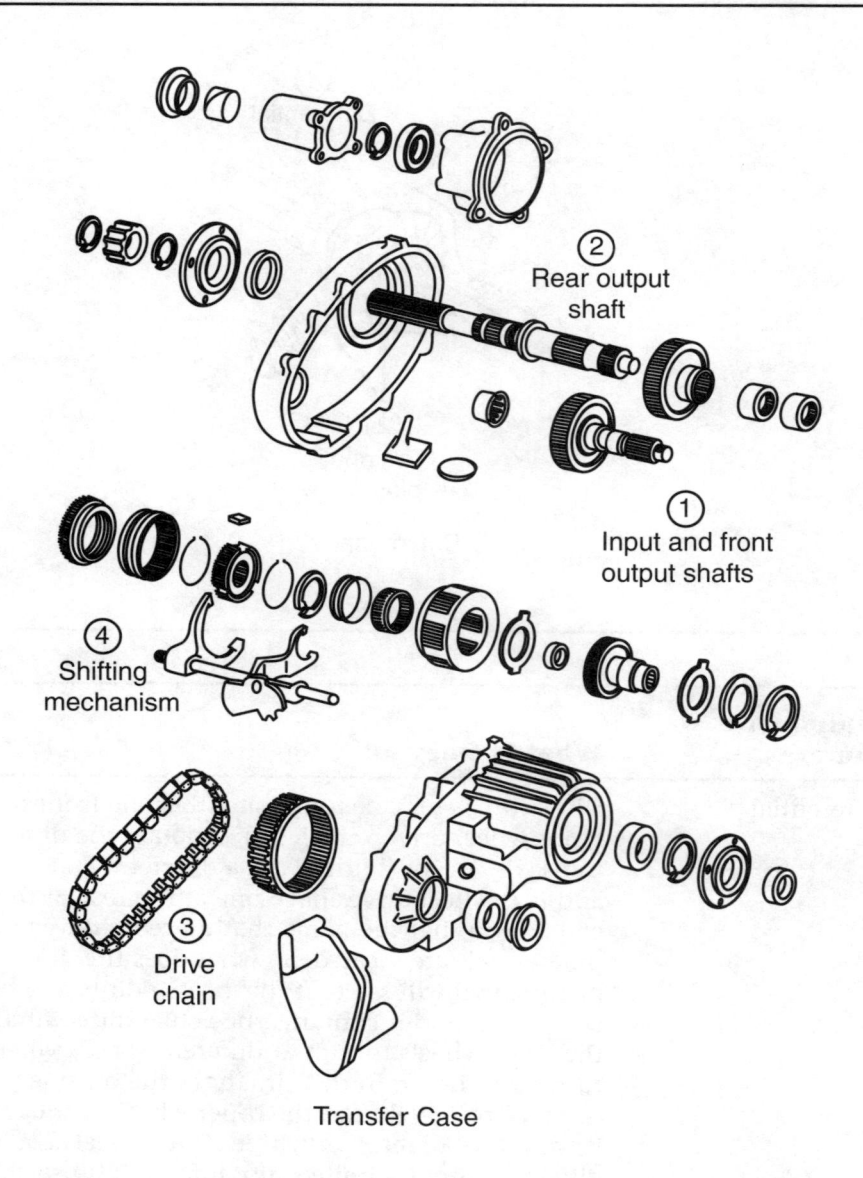

4 WHEEL DRIVE TRANSFER CASE

Diagram Number	Component Name	What It Does
1.	Input and Front Output Shafts	Before you panic about all of the gears present in the drawing, notice that there are only four numbers on the drawing. We want to draw your attention to just a few areas and features of a transfer case. First let's discuss what a transfer case does. Number 1 identifies the input and front output shaft areas. A transfer case is a gearbox that engages or disengages the front and rear axles. There are usually two different ratios inside the box, which we will discuss below. The transfer case is either mounted behind or physically bolted to the rear of the transmission via the input shaft. The front output shaft connects to the front driveshaft and the front axle/differential.

TRANSFER CASE—*Continued*

4 WHEEL DRIVE TRANSFER CASE

Diagram Number	Component Name	What It Does
2.	Rear Output Shaft	The rear output connects to the rear driveshaft and rear axle/differential. This is the output that is used in both 2- and 4-wheel drive. Unless the transfer case is shifted to neutral, this drive shaft always is driven.
3.	Drive Chain	Many transfer cases use a chain to drive the front output shaft off of the main shaft when 4-wheel drive is engaged. Other transfer cases use a gear-to-gear front drive.
4.	Shifting Mechanism	Conventional transfer cases have a shift lever inside the vehicle to allow the drive to select between 2-wheel drive (rear wheels only), 4-wheel low (all 4 wheels driven slower while the engine runs faster; this helps to climb hills or do heavy work), and 4-wheel high (all 4 wheels driven at normal speed). Some transfer cases have a synchronized shift to allow them to be shifted "on the fly." Others have electronic systems that use a motor instead of a shifter to change the gear selection. There many variations on this theme, but these are the most common types. You will be adequately prepared if you understand what the transfer case does.

Task C.9.4   **Provide basic use and installation instructions.**

Use the information in the tables above to help explain the questions your customers may have. Always keep in mind that it is better to recommend them to a shop equipped to repair some of these components than to get involved with an underqualified person.

10. Brakes (3 Questions)

Tasks C.10.1 C.10.2 C.10.3   **Identify major and related components and their function.**

We are going to spend some time looking at the brake system, because it is a system that parts specialists must be well versed in. In the diagram and chart, we are going to look at a typical disc/drum system and a generic anti-lock brake system.

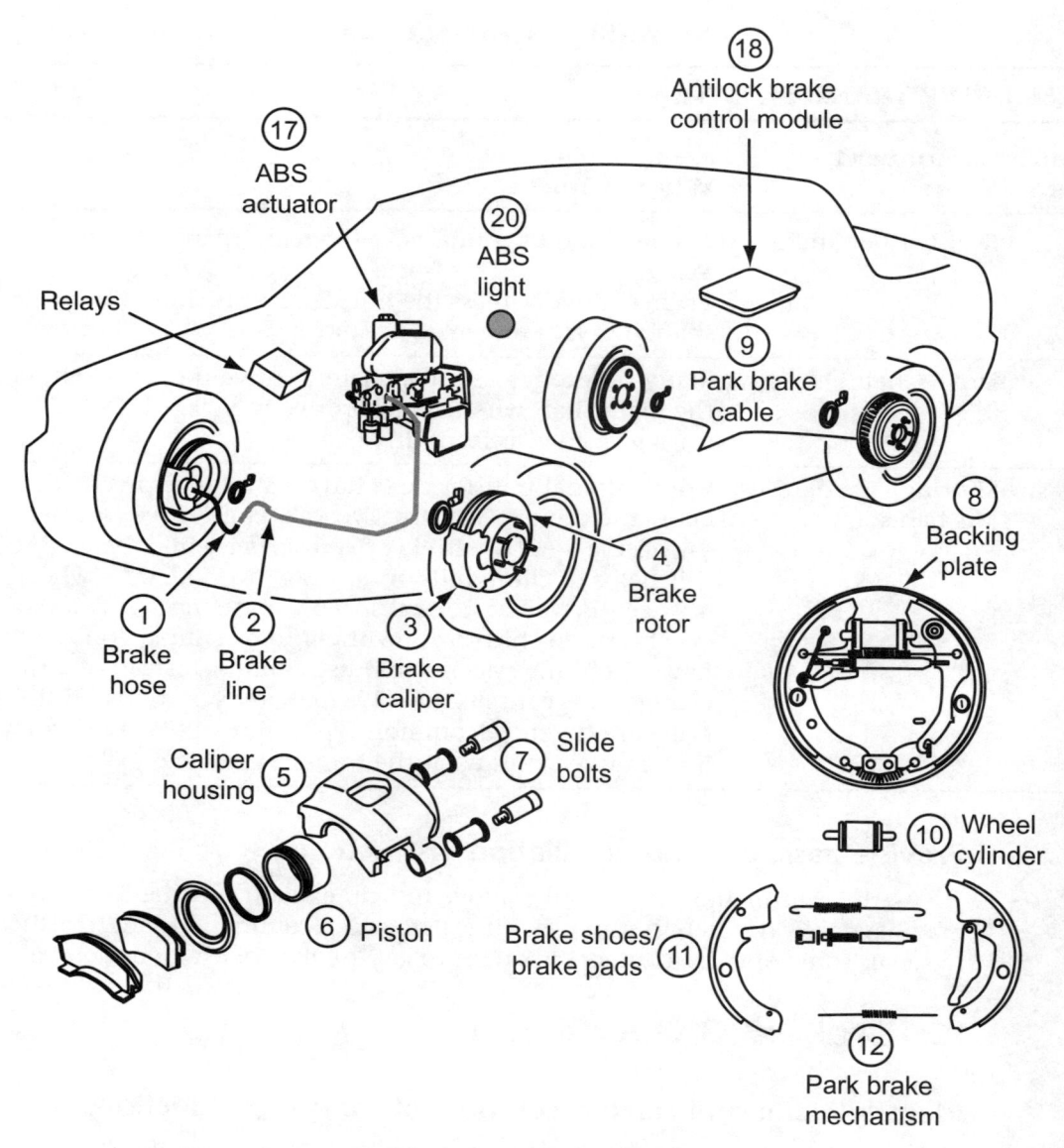

48 Automotive Parts Specialist (Test P2)   Overview of the Task List

Generic Disc/Drum Brake System with ABS

BRAKE SYSTEM

Diagram Number	Component Name	What It Does
1.	Brake Hose	Flexible hoses that connect the brake components at the wheels to the hard brake lines on the body. They allow suspension movement.
2.	Brake Line	Hard lines that are attached to the body of the vehicle and run from the various components, starting at the master cylinder and working out to where the brake hoses connect to them.
3.	Brake Caliper	The Brake Caliper is the moving component of a disc brake. It works like a clamp to grab the spinning rotor and slow the vehicle.

GENERIC DISC/DRUM BRAKE SYSTEM WITH ABS—*Continued*

BRAKE SYSTEM

Diagram Number	Component Name	What It Does
4.	Brake Rotor	This is the namesake of the disc brake system. The rotor has two machined sides that allow the caliper to squeeze. The rotor turns at wheel speed and also acts as a radiator to shed the heat generated by stopping.
5.	Caliper Housing	This is the bare housing component of the caliper. It functions as a mounting bracket—the piston bore is part of it—and often has machined surfaces that allow the assembly to "float" on pins or slides to compensate for wear of the brake pads.
6.	Caliper Piston	The piston is more like a deep round puck. It may be made of steel, aluminum or special plastic compounds. Most brake calipers have only one, but high performance applications and heavy vehicles may have multiple pistons. When the brake pedal is depressed, the master cylinder creates pressure in the system, which causes the piston to push out of its bore toward the brake rotor; the opposing force of the brake pad on the opposite side of the rotor causes a clamping force that stops the vehicle.
7.	Slide Bolts	Many brake systems use bolts that tighten into the spindle. The caliper floats on these. Others may use plates that allow the caliper to slide. These slides are critical to even brake pad wear. If the caliper does not slide, the inner pad often wears out very quickly because the pressure on the rotor does not equalize.
8.	Backing Plate	On drum brake systems, the backing plate is the mounting plate for all of the components of the system.
9.	Park Brake Cable	If you follow the diagram, you will see that the park brake cable connects at one end to the parking brake pedal or lever, and at the other end it attaches to the drum brake by a special cam mechanism that pushes the brake shoes out mechanically to secure the vehicle in a parked position.
10.	Wheel Cylinder	The wheel cylinder performs the same purpose as the brake caliper. Instead of clamping though, it pushes out on the brake shoes.
11.	Brake Shoes/ Brake Pads	The component that is designed to wear out as it does its job, the brake shoe or disc pad has a very complex job. These parts are made up of metal and organic materials that are called friction materials. This is easier to understand if you understand how brakes do their work. Let's agree that there are two types of stops—the slowing stop and the panic stop. The brake pad takes the motion energy of the vehicle and converts it to heat. It does this by dragging its feet like Fred Flintstone on the brake rotor or drum. The rotor and drum provide a smooth surface for maximum heat transfer. The drums and rotors then radiate this heat out to the atmosphere to help keep the brake pads cool enough to do their job. Now back to our two kinds of stops. Brakes must do an amazing job of performing the slow stop, which is a progressive conversion of motion to heat, and then making the panic stop when called upon. This grace under pressure is not how brakes have always worked. Many older vehicles would lock up the brakes under anything more than very gentle application. So, next time you stop your vehicle, take a second to feel how smoothly it can perform each type of function.

GENERIC DISC/DRUM BRAKE SYSTEM WITH ABS—*Continued*

BRAKE SYSTEM

Diagram Number	Component Name	What It Does
12.	Park Brake Mechanism	When the vehicle is parked, some drivers use the park brake to insure it will be there when they come back. The park brake mechanism mechanically actuates the rear brakes in all but a very few applications (like older Subarus). When there are disc brakes on the rear, a similar device mechanically pushes the pads out to clamp the rotor. This handy safety device is almost always linked to the components that automatically adjust the brakes. So, using the park brake helps to adjust the rear brakes. Backing the vehicle up will also actuate most drum brakes self-adjusters.
Not Pictured	Master Cylinder	This is the first component of the hydraulic portion of the brake system. When the brake pedal is depressed, it generates pressure to actuate the brake calipers and wheel cylinders. Pressures can be as high as 2000 psi.
Not Pictured	Power Brake Booster	Brake boosters provide the assistance to push the brake pedal; that is called power brakes. Most use vacuum applied to the master cylinder side of a chamber with a diaphragm in the center to supply this assist. Some systems are known as hydro-boost and use a similar arrangement, with the power steering pump supplying the pressure for assist.
Not Pictured	Pedal Assembly	The pedal is the component that attaches to the master cylinder pushrod and gives the driver a mechanical advantage due to a pedal ratio designed into the mechanism. This makes it so that even without assistance, 10 pounds of pedal pressure may create 100 pounds of brake pressure.
Not Pictured	Combination/ Proportioning Valve	In non-abs vehicles, this component is used to control the amount of pressure sent to the rear brakes. If pressures are equal at all 4 wheels, the rear brakes will lock up first because of the difference in weight. Most vehicles are lighter in the rear than the front. This component usually has a pressure differential valve/switch that will shut down part of the brake system, should a large loss of pressure, like a leak, occur. It will also turn on the red brake light on the dash.
17.	ABS Actuator	An electro hydraulic component, this unit contains the electronic versions of the proportioning valves that are run by the ABS control module to control wheel lock up.
18.	ABS Control Module	This component analyzes inputs from the wheel speed sensors and, sometimes the vehicle speed sensor, to control wheel lock-up. This is very much a simplification because of the vast number of systems of varying complexity out there.
Not Pictured	Brake Warning Lamp	The red brake warning lamp comes on when either the park brake is not fully released, the brake fluid is low, or the pressure differential switch in the combo valve has tripped. In some applications, the ABS system can turn this light on along with its malfunction indicator light.
Not Pictured	ABS Warning Lamp	This light comes on when the ABS system is not functioning. Some systems turn it on when brake fluid is low.

Overview of the Task List — Automotive Parts Specialist (Test P2) — 51

Task C.10.4

Provide basic use and installation instructions.

All brake return springs should be inspected for distortion and stretching. Brake shoes should be cleaned with a shop rag and inspected for broken welds, cracks, wear, and distortion. If the wear pattern on the brake shoes is uneven, the shoes may be distorted. Check all clips and levers for wear and bending. Inspect the brake linings for contamination from oil, grease, or brake fluid. Clean and lubricate adjusting and self-adjusting mechanisms.

The backing plate should be cleaned with an approved brake cleaner that does not allow the release of asbestos dust to the shop air. Inspect the backing plate for distortion, cracks, rust damage, and wear in the brake shoe contact areas. A distorted backing plate may cause brake grabbing. Check the anchor bolt for looseness that may result in brake chatter.

11. Suspension and Steering (3 Questions)

Tasks C.11.1 C.11.2 C.11.3

Identify major and related components and their function.

The suspension is what holds the vehicle up, provides a smooth ride, keeps the tires in contact with the ground, and provides safe control when turning control. We are going to review each of the components in a short-long arm suspension, which is the most common type. The version we selected uses a suspension strut and is typical of both front and rear suspensions of many vehicles. Before we begin, let's quickly cover the two other types of suspensions commonly in use. The leaf spring suspension is found on the rear of many vehicles and the front of heavy-duty pick-ups. In this suspension, a solid axle is mounted with a leaf spring on each side of the vehicle. The spring functions as a control arm, locating the rear end fore and aft in the vehicle in addition to providing normal spring function. More and more vehicles, including trucks are being designed with independent suspension on all 4 wheels, so this design will probably remain only in heavy trucks in the future. The second variant is the McPherson strut. In this design the strut functions as the upper suspension attaching point. These suspensions do not have an upper control arm. The strut provides a pivot point at the top in the form of a strut bearing or mount that allowed the strut to turn with the wheels. The McPherson strut design is being replaced by better handling and riding short-long arm designs. The strut pictured in the diagram is not a McPherson strut because it is not a suspension member. It functions as the shock with an integral coil spring. The short-long arm suspension gets its name from the short upper control arm and long lower control arm that make up its design. This design handles very well because it allows the tire to move in the direction of negative camber as the body rolls going around corners. We will discuss camber when we look at alignment at the end of this section.

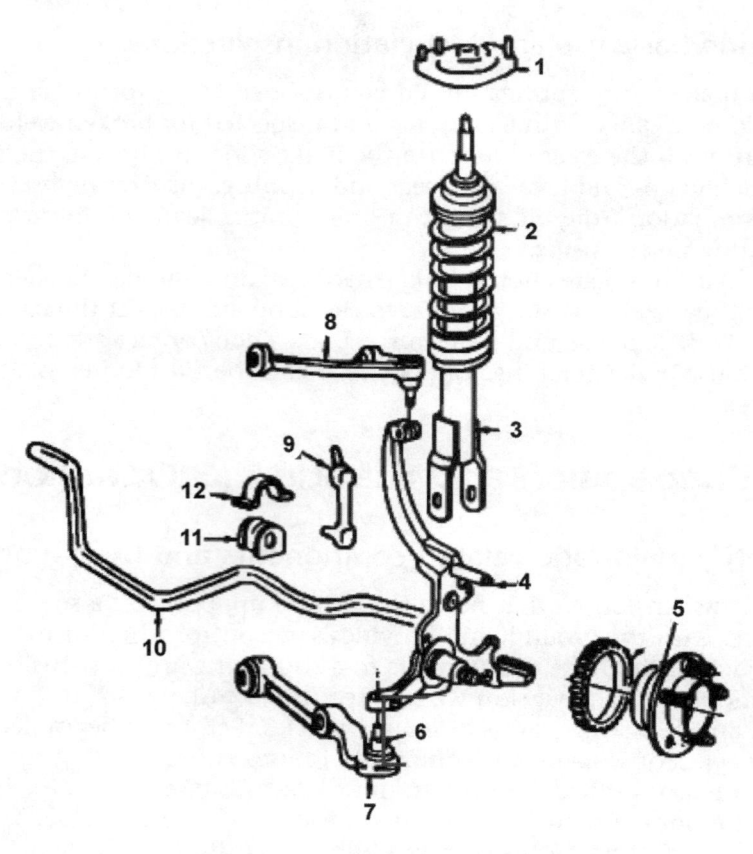

SLA Suspension with Strut

SUSPENSION

Diagram Number	Component Name	What It Does
1.	Strut Mount/ Bearing	The strut mount cushions and isolates the strut assembly from the body. In McPherson applications, it also houses a bearing to allow the strut body to rotate when the wheels turn.
2.	Spring	The coil spring is most common. The previously mentioned leaf spring is used with live axle applications, and one other variation, known as the torsion bar, round out the spring types used in modern vehicles. The torsion bar is a long straight spring that functions by twisting. It is attached at one end to the body and the other end to the suspension. It is usually adjustable to attain correct preload and ride height.

SLA SUSPENSION WITH STRUT—*Continued*

SUSPENSION

Diagram Number	Component Name	What It Does
3.	Strut	The strut is a shock absorber that functions as a mount for the coil spring in some applications. The SLA suspension may use a strut like the diagram, or it may have a shock absorber.
4.	Spindle/Knuckle	The steering knuckle or spindle provides the movement during steering along with the mounting areas that accommodate the wheel hubs, brake components, and the ball joint ends of the control arms.
5.	Hub and Wheel Bearing	This component is shown as a stand-alone piece but may be integrated into a brake rotor with serviceable bearings. The type shown uses a sealed bearing.
6.	Ball Joint	Ball joints are like the joints in our shoulders or hips. They provide for the multi-plane movement that occurs as the vehicle moves up and down and when the wheels turn. They get their name by a ball-shaped bearing surface that rides in lined cup. The ball joint is usually bolted or pressed into the control arm and has a long stud that is bolted into the steering knuckle.
7.	Lower Control Arm	The lower arm is the longer one of the two. It joins the suspension from the steering knuckle to the body of the vehicle. It often has mounting points for shocks, stabilizer bars, and sometimes springs.
8.	Upper Control Arm	The upper control arm is the short arm. It joins the suspension from the steering knuckle to the body of the vehicle. It, too, often has mounting points for shocks springs.
9.	Stabilizer Bar Link	There are several designs, but the stabilizer bar link or end link connects the ends of the stabilizer bar to the suspension.
10.	Stabilizer Bar	Also known as a sway bar, the stabilizer bar is an example of a torsion bar. When a force is applied to one side of it there will be an equal and opposite force on the other end. Therefore, when a vehicle rolls to the outside as it makes a turn, the force on the outside wheel will be transferred to the inside wheel to keep the tire in contact with the ground.
11.	Stabilizer Bar Bushing	To allow the stabilizer bar to pivot on the body of the vehicle, a bushing is located on the body equal distances from the center.
12.	Stabilizer Bar Bracket	This is the bracket that holds the sway bar and bushing to the body of the vehicle.

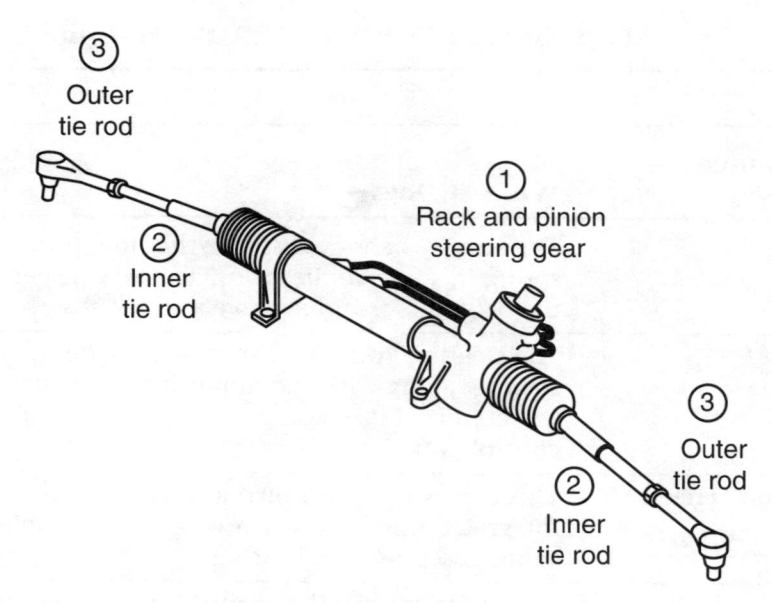

Rack and Pinion Steering

RACK AND PINION

Diagram Number	Component Name	What It Does
1.	Rack and Pinion	Rack and pinion steering gets its name from the internal components. The rack is a long gear that runs horizontal inside the housing. The tie rods attach to the ends of the rack. The pinion is a small gear that connects to the steering column shaft. The reason for using gears is to change the ratio that the wheels turn to steering input. For example, if you could turn the wheels all the way in one rotation of the steering wheel, the vehicle would feel very twitchy and drivers would overcompensate in panic situations. When a rack and pinion is power-assist, a valve is located at the top of the pinion to provide hydraulic assist when there is movement of the pinion.
2.	Inner Tie Rod	Tie rods are like small versions of ball joints. They allow rotational and multi-plan movement to follow suspension movement. The inner socket on the rack and pinion is an open joint because a bellows boot similar to a CV joint boot covers it and protects it from contamination.
3.	Outer Tie Rod	The outer tie rod connects the inner tie rod to the steering knuckle. The two thread together and allow alignment changes to the toe setting by adjusting their lengths.

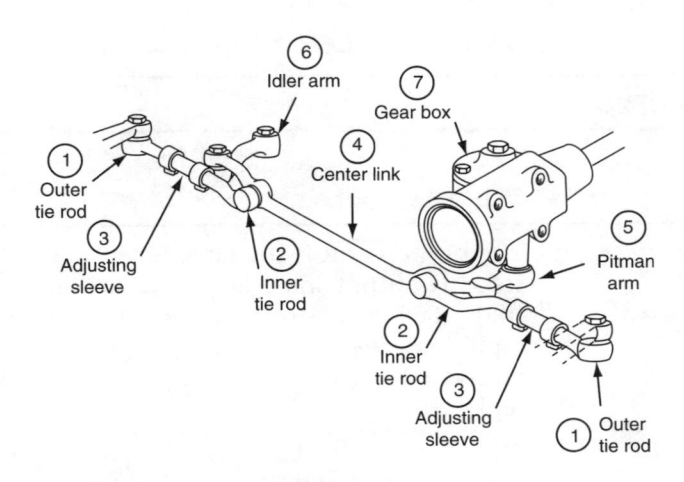

Linkage steering

LINKAGE TYPE STEERING

Diagram Number	Component Name	What It Does
1.	Outer Tie Rod	Tie rods are like small versions of ball joints. They allow rotational and multi-plane movement to follow suspension movement. The outer tie rod connects the inner tie rod to the steering knuckle. The two thread together and allow alignment changes to the toe setting by adjusting their lengths. The components of the suspension must move up and down at the wheels. The components of the steering must allow this movement while causing the least amount of change in the toe settings of the alignment. Tie rod ends allow around 30 to 60° of movement.
2.	Inner Tie Rod	The inner tie rod works just like the outer tie rod. In most cases there is an adjusting sleeve between the two. The inner tie rods may attach to a center link, to the tie rod for the other side of the vehicle, or directly to the steering box. Many inner and outer tie rods are threaded with one left- and the other right-handed so that when the technician adjusts the tie rods, they rotate in the same direction to adjust toe in or out on each side of the vehicle.
3.	Adjusting Sleeve	The sleeves are generally made of steel and have clamps at each end where the tie rods thread into them. After the technician adjusts the toe to specification, he tightens the clamps on the adjusting sleeve to lock the toe setting. When one or both tie rods are replaced, the adjusting sleeves are often rusted and frozen necessitating their replacement. Often they are made with right- and left-handed threads, so it is important to assemble and install them correctly.
4.	Center Link	The center link is almost always used in conjunction with an idler arm. The center link attaches between the steering box and idler arm and maintains a parallel plane across the front suspension. In most center link applications, the tie rods attach to the center link and then to the steering knuckles. A similar variation of this theme is the drag link. The drag link generally attaches from the steering gear to a spindle. Like the center link, it does not provide for toe adjustment but may be adjustable to allow the steering box or wheel to be centered.

LINKAGE STEERING—Continued

LINKAGE TYPE STEERING

Diagram Number	Component Name	What It Does
5.	Pitman Arm	The pitman arm is the large lever that bolts to the bottom of the steering output shaft. Some pitmans may have ball joint–like joints that attach them to the rest of steering linkage. These require replacement when they wear. The other type of pitman has a hole to receive the joint of a tie rod and usually does not need to be replaced.
6.	Idler Arm	The idler arm is located on the passenger side of the vehicle in all but a very few applications that use two idler arms. The idler arm is responsible for providing mirror image movement of the idler arm or its opposing idler arm to keep the center link in precise parallel alignment. The bushings in the idler arm are the parts that generally wear resulting in a wandering sensation on the road.
7.	Steering Gear	The steering gear may be power-assisted or manual. It has a gear arrangement that uses a set of ball bearings that run in a cage between the gears to provide smooth steering operation and minimum reaction back to the driver when going over bumps. This design is known as recirculating ball steering and is the most prevalent design used in the last 30 years. Common failures are usually leaks at output or input shaft seals and leaking power steering line connections. The mechanical part of steering boxes is highly dependable due to the importance of its function.

Task C.11.4 **Provide basic use and installation instructions.**

Probably the single most important area for you to advise your customer is safety. The force stored in one coil spring can do considerable damage to the vehicle and your customer if proper precautions are not taken.

12. Heating and Air Conditioning (3 Questions)

Tasks C.12.1 C.12.2 C.12.3 **Identify major and related components and their function.**

The heating and air conditioning systems provide comfortable air inside the cabin of the vehicle. Warm air is generated using the heat created by the engine. The conductor for this is known at the heater core. The heater core is a small radiator. Coolant from the engine is circulated through it and a blower blows across the core to provide hot air inside the vehicle. There are several passages for the air to flow through that are controlled by the controls on the dash. The controls may have cables, vacuum actuators, or electric motors that run a series of doors inside the ducts to direct the air where the driver wants it. Temperature is controlled by a blend door and in some cases a water control valve that limits the amount of warm engine coolant that enters the heater core.

Air conditioning works through the same set of ducts, but because there is no naturally cool fluid on a vehicle it must create coldness. This is done by evaporation and condensation. The air conditioning system uses a compound that has a very low boiling point. This compound, simply called R-134A, is a refrigerant. The air conditioning system takes advantage of the refrigerants low boiling point to move it through two states—liquid and gas. Follow the diagram below to see how this occurs. Think of how water becomes steam when it boils and you will understand. Also think of how a wet rag swung in the air fast becomes cold—this is called evaporation.

So we have a compressor that has both suction and discharge sides to it. It creates movement and pressure within the system. The compressor pushes gaseous refrigerant to the condenser, which sits in front of the radiator. As the gas passes through the tubes of the condenser, it changes to a liquid and becomes colder. The compressor pumps it toward the evaporator inside the vehicle. Between the compressor and the evaporator, there is a restriction. This can be in the form of a control valve or just a small tube called an orifice. This causes the liquid refrigerant to spray into the evaporator. At this point, the pressure on the liquid drops, and it rapidly becomes a gas. This process is called evaporation. Think of our wet rag again. This causes the evaporator to become very cold so that a fan blowing across it will yield cold air. The compressor starts the whole cycle over again. The components of the air conditioning system share most of the ducting with the heater but have electrical controls to engage the compressor. The system may have switches to protect the compressor from extremely low or high pressures. Many air conditioning systems are monitored and controlled by a body control computer or the PCM.

Task C.12.4 Provide basic use and installation instructions.

The majority of your HVAC customers will be professionals, due to the type of equipment necessary to service A/C systems. As with automatic transmissions, identification of systems is always the challenge. The retrofit component of A/C service has been a moving target with regard to the manufacturers' recommended procedures. Helping your customer stay up to speed by providing information you receive from your manufacturers.

13. Electrical Systems (2 Questions)

Task C.13.1 Identify major components.

The electrical system is any of the systems and subsystems that make up the automobile wiring harnesses, such as the lighting system or starting and charging system.

Vehicles have many circuits that carry electrical current from the battery to the individual components. The total electrical system includes subsystems such as the starting system, charging system, lighting system, and accessory systems.

The starting system initiates engine operation. When the ignition key is turned to the START position, a small current flows from the battery to a solenoid or relay switch which, in turn, closes another electrical circuit that allows full battery voltage to the starter motor. The starter motor then turns the flywheel, mounted on the rear of the crankshaft, to start the engine and put all engine parts in motion.

The charging system maintains the battery's state of charge and provides power for all of the electrical systems and accessories. In addition to the battery, the charging system includes the engine-driven alternator (or generator), voltage regulator, dash gauge (or indicator light), and associated wiring.

Task C.13.2 Task Identify component function.

The starting system operates as follows: When the ignition key is turned to the START position, electric current is sent to the solenoid, and battery voltage is supplied directly to the starter motor. The starter motor then turns a flywheel mounted on the rear of the crankshaft that starts all engine parts in motion. The ignition system provides a spark to the spark plugs which ignite the air/fuel mixture from the carburetor. If all components are in good working condition, the engine should start immediately.

The charging system performs two basic tasks: maintaining the battery's state of charge and providing electrical power for the vehicle while it is running. Turned by a drive belt that is driven by the engine crankshaft, an generator converts mechanical energy into electrical energy. When the electrical current flows into the battery, the battery is said to be charging. When the current flows out of the battery, the battery is said to be discharging.

The vehicle must be protected from fire hazards that could occur if a powered circuit is accidentally shorted or grounded. Fuses, circuit breakers, and fusible links may be used to protect circuits.

Task C.13.3 **Identify related items.**

Related items to electrical are every system on the modern vehicle. You will find that sometimes you must completely switch catalogs when locating components because of the number of electrical components found in every system.

Task C.13.4 **Provide basic use and installation instructions.**

When we go through the tests and explanations we will provide you with opportunity to gain an understanding of what it means to provide basic use and installation instructions.

14. Battery, Charging, and Starting Systems (3 Questions)

Task C.14.1 **Identify major components.**

The electrical/charging system performs two basic duties, maintaining the battery's state of charge and providing electrical power for the ignition system, air conditioner, heater lighting and all electrical accessories. Electrical energy is stored in the battery. This energy is used to supply the power needed to start the motor and other devices on the vehicle. The charging system is responsible for restoring power back to the battery through electrical output from the alternator. There are several components of the electrical/charging system. These components must operate together. These components include: battery, alternator, starter, starter relay, ignition switch, power module, serpentine belt, and coil. Changes in the electrical system over the past 20 years has seen the introduction of fully electronically controlled engines which eliminate the distributor with a set of coils that are triggered by sensors which read timing marks on the flywheel. As the electrical systems become more and more complex, the wiring harness continues to get fatter and heaver. Currently, the control units or computers are being linked together. A control module will be placed at each electrical component picking up signals from a single control wire that will run around the car. There are a limitless number of ways electrical systems can be designed but each design must consist of the following three components.

1. There must be a source of electricity. (Battery and the charging system which is a subsystem of the electrical system)
2. There must be a load. The load is the business end of the circuit.
3. There must be a circuit connecting the source to the load.

When selling electrical components, your main concern will be in determining that the circuits are complete, sources are adequate (battery), circuits are designed appropriately for the task to be performed, and load devices are functioning properly.

Task C.14.2 **Identify component function.**

The following components are in all charging systems:

Alternator—Converts mechanical energy into electrical. Produces the voltage needed to supply the electrical energy needed once the automobile is running. The alternator also recharges the battery once the car has started.

Battery—Stores electrical current released by the alternator and is used to start the automobile. The battery also supplies voltage to the vehicles electrical accessories when the engine is not running.

Fusible Link—Short wire sections of lighter gauge with special insulation that puffs up when the fuse is blown.

Relay—The relay located in the starter circuit is a magnetic switch used to close the ignition switch in the START position allowing voltage to flow to the battery through the relay, to the starter windings, and back to the battery by a common ground.

Solenoid—A pull in device that is energized when the drives close the ignition switch to complete the cranking circuit.

Starter—High torque, electrical device that accepts energy from the battery and converts it into mechanical force to crank the engine.

Voltage Regulator—Controls the amount of voltage produced and sent to charge the battery.

Task C.14.3 Identify related items.

Some related items that are of interest when it comes to the charging system are as follows:

Cold Cranking Amps (CCA)—Number of amps the battery can deliver at zero degrees to start the engine.

Reserve Amps—Output of amperage while there is no charging input. For example: the number of minutes the battery is able to maintain a charge when leaving the lights on after turning the engine off.

Wiring Harness—A conductor used to pass electrical current throughout the automotive system. Wire is specified by gauge.

Serpentine Belt—The rubber belt used to spin the alternator which generates the charge for the battery once the vehicle is operational.

Task C.14.4 Provide basic use and installation instructions.

Caution: Always wear safety goggles or glasses when working with batteries no matter how small the job. Replacing the battery is easily completed by first insuring the vehicle engine is off. Disconnect the battery negative terminal clamp, use a terminal clamp puller to remove the negative cable, loosen the battery positive cable, use a terminal clamp puller to remove the positive clamp, remove the battery hold down hardware and any heat shields, remove the battery from the tray, at this point you may install a new battery by, placing the new battery into the battery tray, install the battery hold down hardware and connect the positive battery cable, then the negative battery cable, then coat the terminals with corrosion preventative spray. Check installation by starting the car.

Alternator replacement is unique by make and model but there are some inspections you should complete before you decide it is time to replace the alternator. Check the belt for tension and proper pulley alignment. Inspect the belt, the battery, its posts and cable connections, all system wiring and connections, and the alternator for loose or missing bolts. Once you have determined that the cause of the problem is your alternator, replace it. Refer to the owner's manual for specific steps.

15. Miscellaneous (3 Questions)

Task C.15.1 Identify fastener thread types (SAE, USS, and metric).

Three fastener types commonly used in the automotive trade-the United States Standard (USS), the American National Standard (ANS), and the Society of Automotive Engineers Standard (SAE)-have all been replaced by the Unified National Series. The Unified National Series consists of four basic classifications: (1) Unified National Coarse (UNC or NC), (2) Unified National Fine (UNF or NF) (SAE), (3) Unified National Extra Fine (UNEF or NEF), and (4) Unified National Pipe Thread (UNPT or NFIT).

Task C.15.2 **Identify fastener thread diameter and pitch.**

The two common metric threads are coarse and fine and can be identified by the letters SI (Systeme International d'Unites or International System of Units) or ISO (International Standards Organization).

To identify the type of threads on a bolt, bolt terminology must be defined. The bolt has several parts. The bolt head is used to torque or tighten the bolt. A socket or wrench fits over the head, which enables the bolts to be tightened. Common U.S. Customary (USC) and metric sizes for bolt heads include sizes given in fractions of an inch (English) and in millimeters (metric).

Task C.15.3 **Identify fastener type.**

Although some USC and metric head sizes are very close, never use a metric wrench or socket for USC bolts, or vice versa. Tool slippage may cause injury or damage the bolt head.

Bolt diameter is the measurement across the major diameter of the threaded area or across the bolt shank. The thread pitch of a bolt in the English system is determined by the number of threads there are in one inch of threaded bolt length and is expressed in number of threads per inch. The thread pitch in the metric system is determined by the distance in millimeters between two adjacent threads. To check the thread pitch of a bolt or stud, a thread pitch gauge is used. Gauges are available in both English and metric dimensions. Bolt length is the distance measured from the bottom of the head to the tip of the bolt. The bolt's tensile strength, or grade, is the amount of stress or stretch it is able to withstand. The type of bolt material and the diameter of the bolt determine its tensile strength. In the English system, the tensile strength of a bolt is identified by the number of radial lines (grade marks) on the bolt head. More lines mean higher tensile strength. In the metric system, tensile strength of a bolt or stud can be identified by a property class number on the bolt head. The higher the number, the greater the tensile strength.

Task C.15.4 **Identify fastener grade.**

It is very important to be familiar with the standard bolt indication measurements and grade markings. All bolts in the same connection must be of the same grade. Otherwise, they will not perform equally. Likewise, nuts are graded to match their respective bolts. For example, a grade 8 nut must go with a grade 8 bolt. If a grade 5 nut is used instead, a grade 5 connection will result. The grade 5 nut cannot carry the loads expected of the grade 8 bolt. Grade 8 and critical applications require the use of fully hardened flat washers. They do not dish out like soft wrought washers that cause loss of clamp load.

Bolt heads can pop off because of fillet damage. The fillet is the smooth curve where the shank flows into the bolt head. Scratches in this area introduce stress to the bolt head, causing failure. The bolt head can be protected by removing any burrs around the edges of holes. Also, place flat washers with their rounded, punched side against the bolt head and their sharp side to the work surface.

Task C.15.5 **Identify fitting type.**

Flare, compression, and pipe are three types of fittings. A flare fitting requires that the end of the tubing be expanded at an angle, or flared. This operation requires the use of a flaring tool. Compression fittings do not require any special tools. A soft metal band, known as a ferrule, is placed between the end of the tubing and the fitting. As the two fittings are tightened together, the ferrule is crimped into a sealing connection between the fittings and tubing. Pipe fittings seal using tapered threads. The more that pipe fittings are tightened, the closer the threads are brought together.

Task C.15.6 **Identify fitting size.**

Fittings should be manufactured to meet the functional requirements of the Society of Automotive Engineers (SAE), Automotive Service Association (ASA), and the American Society of Mechanical Engineers (ASME) for low, medium, or high pressure line connection service as applicable.

Pipe fittings made of brass or iron can be used with air, water, oil, fuels, and various gases. They may be connected by brass, copper, or iron pipe. These fittings are applied in a wide variety of automotive applications including off-road as well as on-road equipment. Available styles include nipple, tee, elbow, adapter, plug, and cap. Male and female pipe fittings are available in popular sizes from ⅛ inch to 2 inches.

Flare fitting applications include refrigeration equipment, air compressors, oil burners, and most any type of machinery. Flare fittings may be used with thin-wall tubing where application requires joints to withstand high pressures (up to 2,800 psi [19,306 kPa]) using 0.030-inch [0.762-mm] wall copper tubing). They may also be used with flareable copper, brass, aluminum, and welded steel hydraulic tubing. Tubing flared at 37 or 45 degrees (as applicable) forms a sound joint resisting mechanical pull-out. It seals and remains leak-free even when disconnected and reconnected. Flare fittings are available in long and short nuts, plugs, unions, elbows, tees, and adapters. They are available in male, female, or male-to-female configuration. Popular flare fitting sizes are from ¼ inch to ⅝ inch.

Compression fittings eliminate the need to flare, solder, or otherwise prepare tubing before assembly. They are for use where excessive vibration or tube movement is not a problem. There are some applications, such as brake service, where compression fittings are not used. Compression fittings are used with copper, brass, or aluminum and are easily installed. All sealing surfaces are precision machined to ensure reliable and leakproof closures. Compression fittings include sleeve(s) and nut(s). They are available in many configurations including unions, elbows, and tees, in popular sizes of ¼ inch to ⅝ inch.

It should be noted that many variations of fitting combination types are available. These include, but are not limited to, flare to pipe, flare to compression, and compression to pipe. These include male-to-female as well as female-to-male provisions. Sizes given are for English. However, many are available in metric sizes as well.

Task C.15.7 **Identify body repair and refinishing materials and supplies.**

Automotive refinishing materials and procedures have changed considerably in recent years. There are many brands and types of automotive paints now available. As a result, the auto body technician must learn new technologies to be able to apply the various types of automotive finishes.

The terms *paint* and *finish* are generally used to describe many different finishing materials. They are, however, most commonly used when referring to the topcoat or outside layer (about 0.004 inch [0.102 mm] thick) applied to metal or plastic components.

All paints contain three basic ingredients: binders, solvents, and pigments. Special additives are also used in some paints to alter their properties and characteristics. Binders are the resinous, film-forming ingredients that adhere to the substrate, the surface being painted. Pigments are added to the paint to give it color. The amount and type of pigment will somewhat alter the durability, adhesion, flow, and other characteristics of the paint. A number of additives are used in paint to alter its properties and characteristics. Manufacturers' instructions should always be followed when mixing additives with paint.

Primers are used to help improve paint adhesion on a metal or plastic substrate. The general term *primer*, also called prepcoat, implies that the substrate is being prepared for a final coat.

Solvents, also called vehicles, are added to paint so that it can be applied easily. Solvents thin the paint pigments and binders so that they may be sprayed. Solvents are sometimes called volatiles because they vaporize and evaporate easily and rapidly.

Lacquer topcoats dry from the outside in as their solvents evaporate. Chemical hardeners, or isocyanates, are not used with lacquer finishes. One reason that lacquer finishes are so popular is because they dry rapidly. Lacquer remains more or less soluble, allowing a new coat of acrylic lacquer to bond or unite with an old lacquer finish.

In prior years, original equipment manufacturers (OEMs) used a specially formulated lacquer which was heated (baked) in high-temperature ovens to promote drying and curing. These lacquer-based finishes have now been phased out and have been replaced by various types of enamels, such as the following:

a. Alkyd enamel. A favorite for over forty years, alkyd enamel is the least expensive of the enamels. It usually covers in two coats and is sometimes referred to as a synthetic enamel because the alkyd resin is made from petroleum rather than natural sources.
b. Acrylic enamel. An enamel that consists of a solvent blend of binder and pigment materials. They flow out well to fill small imperfections in the substrate. Acrylic enamels must be applied in a totally dust-free environment and permitted a long time to dry.
c. Polyurethane enamel. An enamel used for automotive refinishing that provides a hard, tile-like, high-gloss finish. It has excellent flow-out and appearance, good adhesion, and flexibility. These finishes have been used on aircraft, off-road equipment, and fleet trucks, as well as automobiles. Polyurethane enamels produce a "wet-look" finish.
d. Acrylic-polyurethane enamel. An enamel that weathers well and provides a higher gloss and greater durability than other polyurethane enamels. A two-part finish, it must be applied immediately after being mixed, which activates it.
e. Waterborne acrylic enamel. An enamel used on some vehicles. This finish is baked at more than 300° F (149° C). After baking, the finish is covered with a polyurethane clearcoat. This process is not practical for repair shops since most do not have baking ovens.

Task C.15.8

Identify hose and tubing types and applications.

The primary purpose of tubing or hose is to transmit fluid, often under high pressure, from one point to another. The power steering pump, for example, moves power steering fluid, via hoses, from the pump to the steering gearbox and returns the fluid, under low pressure, to the pump reservoir. Hoses also function as additional reservoirs and act as sound and vibration dampers.

Hoses are generally a reinforced synthetic rubber material coupled to metal tubing at connecting points. The pressure side must be able to handle pressures up to 1,500 psi (10,343 kPa). For that reason, wherever there is a metal-to-rubber tubing connection, the connection is crimped. Pressure hoses are also subject to surges in pressure and pulsations from the pump. Their reinforced construction permits hoses to expand slightly and absorb changes in pressure.

Where two diameters of hose are used on the pressure side, the larger diameter, or pressure, hose is at the pump end. It acts as a reservoir and as an accumulator absorbing pulsations. The smaller diameter, or return, hose reduces the effects of kickback from the gear itself. By restricting fluid flow, it also maintains constant back pressure on the pump, which reduces pump noise. If the hose is of one diameter, the gearbox is performing the damping functions internally.

Because of working fluid temperature and adjacent engine temperatures, these hoses must be able to withstand temperatures up to 300° F (149° C). Due to various weather conditions, they must also tolerate subzero temperatures as well. Hose material is specially formulated to resist breakdown or deterioration due to oil or temperature conditions.

Caution: Hoses must be carefully routed away from hot engine manifolds. Power-steering fluid is very flammable. If it comes in contact with hot engine parts, it could start a fire.

Steel tubing and flexible synthetic rubber hosing serve as the arteries and veins of the hydraulic brake system. These brake lines transmit brake fluid pressure from the master cylinder to the wheel cylinders and calipers of the drum and disc brakes.

Fluid transfer from the driver-actuated master cylinder is usually routed through one or more valves and then into the steel tubing and hoses. The design of the brake lines offers quick fluid transfer response with very little friction loss. Engineering and installing the brake lines so they do not wrap around sharp curves is very important in maintaining this good fluid transfer.

Task C.15.9 Determine hose and tubing size.

The size and composition of hose depends largely on its application. Brake line tubing, for example, usually consists of copper-fused double-wall steel tubing in diameters ranging from ⅛ to ⅜ inch (3.175 to 9.525 mm). Some original equipment manufacturer (OEM) brake tubing is manufactured with soft steel strips sheathed with copper. The strips are rolled into a double-wall assembly and then bonded in a furnace at extremely high temperatures. Corrosion protection is often added by tin-plating the tubing.

Assorted fittings are used to connect steel tubing to junction blocks or other tubing sections. The most common fitting is the double or inverted flare. Double flaring is important to maintain the strength and safety of the system. Single flare or sleeve compression fittings may not hold up in the rigorous operating environment of a standard vehicle brake system.

Fittings are constructed of steel or brass. The 37 degree inverted flare or standard flare fitting is the most commonly used coupling. Late model vehicles may use ISO or metric bubble flare fittings.

Never change the style of fitting being used on the vehicle. Replace ISO fittings only with ISO fittings. Replace standard fittings with standard fittings. The metal composition of the fittings must also match exactly. Using an aluminum-alloy fitting with steel tubing may provide a good initial seal, but the dissimilar metals create a corrosion cell that eats away the metal and reduces the connection service life.

Task C.15.10 Recommend proper application and usage of chemicals/appearance products.

There are many potentially dangerous materials that are encountered in the automotive body shop; materials that can cause bodily harm and property damage if improperly handled. It is important to always follow the manufacturers' suggestions when working with any such material.

Hazardous materials include any material that can cause physical harm or pose a risk to the environment. Hazardous substances are a subset of hazardous materials. These substances pose a threat to waterways and the environment.

The terms *hazardous materials* and *hazardous substances* have specific legal meanings, as identified and regulated by the United States Environmental Protection Agency (EPA). There are laws designed to protect life and the environment from the careless use of hazardous materials. Misuse can result in prosecution and carry severe penalties. It is important to check state and local laws governing the use and disposal of hazardous materials.

There are four basic types of hazardous materials found in the automotive body shop:

a. Flammable materials-materials that easily catch on fire or explode.
b. Corrosive materials-materials that will dissolve metals and can damage the skin.
c. Reactive materials-materials that become unstable and are likely to burn, explode, and/or give off toxic fumes.
d. Toxic materials-materials that cause illness or death from contact, ingestion, and/or inhalation.

Special compounds are available for absorbing oil and cleaning oil spots. Personnel and customers can slip and fall on dirty, oily, or greasy floors. Due to the many hazards encountered in an automotive body shop, signs should be posted to inform visitors and customers that the work areas are for personnel only.

Electrical outlets should be provided throughout the shop, so that there is no need to have extension cords running across the floor.

There is always the possibility of cuts, scrapes, bruises, and pulled muscles in the auto body shop because the technician often works with sharp, heavy sheet metal tools and parts. Gloves should be worn to protect the hands. When lifting heavy items, keep the back straight and lift with the legs.

Everyone in the shop should know how to administer basic first aid. All accidents and injuries must be reported in accordance with posted shop rules and regulations. Emergency numbers should be posted near each telephone in the shop.

Task C.15.11 Recommend proper application and usage of vision and safety products.

Safety is everybody's responsibility. The use of safety equipment when handling chemicals, aerosols, lubricants, and automotive parts is mandatory. Battery acid, chemicals from paints to bug remover should be handled with care, safety equipment should be worn such as goggles, gloves, long sleeves and boots as a minimum.

Safety can include being aware of the work environment with respect to possible danger. Items such as clothing, hair and jewelry can create times of danger as they can easily become entangled in moving machinery. Some locations may require wearing coveralls to protect employees from these dangers. Hearing protection is also a good idea when working abound high noise areas. And last but not least, lifting and carrying injuries can be avoided by wearing appropriate support equipment and learning to lift heavy objects correctly.

D. Vehicle Identification (3 Questions)

Task D.1 Locate and utilize vehicle ID number (VIN).

The standard location for the vehicle identification number (VIN) is attached to the driver side of the instrument panel and is visible through the windshield.

Task D.2 Locate production date.

The production date is the date which the vehicle was assembled. The standard location for this information is on a tag affixed to the driver's door by the door latch, or on the driver's door sill plate.

Task D.3 Locate and utilize component identification data.

Component identification (ID) data are stamped in the casting, or on a tag that is attached to the component. All major components on the vehicle will have ID data attached.

Task D.4 Identify body styles.

The body styles are as follows:
a. Coupe-a two door vehicle.
b. Sedan-a four door vehicle.
c. Station wagon-a vehicle having a passenger compartment which extends to the back of the vehicle.
d. Regular cab-a cab that has one seating surface and two doors.
e. Extended cab-a cab, larger than a regular cab, having one and a half seating surfaces. Extended cabs may have two, three, or four doors.
f. Crew cab-a cab which has two full seating surfaces and four doors.

g. Step-side bed-a bed that has steps on the outside of the bed and is available in six- and eight-foot lengths.
h. Style-side bed-a bed having no steps available in six- and eight-foot lengths.

Task D.5 **Utilize additional reference material for interpreting component information.**

Vehicle build sheets can be a good source of additional information. Vehicle build sheets include the vehicle's identification number (VIN), color, engine size, transmission type, and axle ratio.

Task D.6 **Locate paint code(s).**

The paint code is not part of the vehicle identification number (VIN). It is located on a separate trim tag, called the service parts identification label, which is mounted under the hood or on the driver's door of some vehicles.

E. Cataloging Skills (7 Questions)

Task E.1 **Locate proper catalog and identify needed part(s).**

Each catalog has, in the upper right-hand corner, the form number, the date of issue, and the number and date of the catalog it replaces or supplements. The cover also identifies the product line and manufacturer and often provides an area where the jobber can stamp his or her name for the convenience of dealer customers. A parts person cannot identify a catalog by size simply because the parts suppliers try to make all catalogs the same size so they fit into a standard catalog rack. Some manufacturers, however, may print their catalogs in odd or different colors so they stand out in the catalog rack.

Task E.2 **Obtain and interpret additional information (footnote, illustration, etc.).**

Every parts specialist will have occasion to refer to catalogs at some point, either to reference additional information, determine exceptions, or find alternative parts.

Sometimes footnotes are difficult to understand when a number of them are squeezed on a page. The footnotes themselves are important. Footnotes should be referenced systematically in order to fully define those applications that have multiple possibilities.

Task E.3 **Utilize additional reference material (technical bulletins, interchange list, supplements, etc.).**

Bulletins are lists of updated information that manufacturers send to jobbers between issues of their catalogs. They are the primary tool for conducting catalog maintenance, which involves continually updating and revising the catalog racks and making sure counter personnel are using the most up-to-date information from those catalogs.

Bulletins are usually one of the following types:
a. New item availability bulletins which list items now in stock.
b. Supersession bulletins, which note part numbers that now supersede previously noted numbers.
c. Product information bulletins, which contain specific information about a product, such as a manufacturing defect or a unique method of installation.
d. Technical bulletins, which alert counter personnel to any unusual installation or fit problems or unique maintenance tips.
e. Correction bulletins, which refer to catalog errors due to mistyping or inaccurately assigned parts numbers.

Task E.4 Identify catalog terminology and abbreviations.

Manufacturers usually include additional aids for using their publications such as an abbreviation list which can make for efficient use of the catalog. For example, abbreviations can often have more than one meaning. FWD can mean front-wheel drive or four-wheel drive. OD can mean overdrive or outside diameter. The definitions of these, and other abbreviations, can only be determined by checking the abbreviation list provided.

Task E.5 Locate index and table of contents.

The Table of Contents will show the catalog sequence and is normally located at the beginning of each catalog. The Table of Contents is used to see where similar information is placed within the catalog. The Index lists the contents and gives reference to page numbers or placement within the catalog. The index is valuable for finding specific information.

Task E.6 Perform catalog maintenance.

Bulletins are the primary tool for conducting catalog maintenance, which involves continually updating and revising the catalog racks and making sure counter personnel are using the most up-to-date information from those catalogs.

F. Inventory Management (2 Questions)

Task F.1 Report lost sales.

The main benefit of using lost sales reports is the ability to spot trends in the aftermarket parts business. These reports will show new part numbers that the store does not stock as well as older part numbers that are increasingly in demand. Some part numbers are assured of almost instant demand, such as those found in some of the newer downside models. Others parts will grow in demand, but somewhat more slowly. The store owner or buyer must analyze not only the part numbers that are selling, but also those that the store does not stock.

Task F.2 Verify incoming and outgoing merchandise.

As with any delivery, an order for a shipment must be carefully picked, packaged, and documented. A packing list or slip must be included with each shipment. This list contains a detailed description of the items included in the shipment. A separate packing list can be placed in each part of a multiple-part shipment, but often a single comprehensive list is packaged with or secured to the first part of a multiple-part shipment.

Task F.3 Perform physical inventory.

Physical inventory should be done once a year. Physical inventory is when all stock is physically counted on the shelves and checked to determine what is actually in the store or shop versus what is in the computer. This is one way to keep the computer system up-to-date and can help you to give the customer faster and more accurate service.

Task F.4 Report inventory discrepancies.

After the physical inventory has been completed, the discrepancies must be recorded and put into the computer system. The discrepancies are reported by filling out an inventory discrepancy form. After discrepancies have been entered into the computer system, they should be kept and filed for future reference.

Task F.5 Perform stock rotation.

Stock rotation is moving the older stock in front of the newer stock. This prevents the older stock from sitting at the back of the shelf and eventually getting an aged appearance. The stock should be rotated every time that the shelves are stocked. Rotating the product so that the labels are toward the front of the shelf is known as facing. Rearranging the shelves for a neater appearance is done whenever necessary to increase appeal to the customer.

Task F.6 Handle special orders.

A special order is placed whenever a customer purchases an item not kept in stock. A part that repeatedly appears on lost sales or special order reports should be considered for stocking status.

Task F.7 Perform proper core handling (i.e.: accepting or declining cores, storage, and return).

The parts counterperson should be careful when handling cores and accepting core returns from customers. A core charge is a charge that is added when the customer buys a remanufactured or reconditioned part. Core chargers are refunded to the customer when the old defective but rebuildable part is returned. To ensure that customers return their cores, a parts specialist should always require a core charge. There are some parts specialists, however, who believe that charging for the core damages customer relations. While some customers might reliably return a core of good value, there are those who do not return any cores or who return cores of little or no value. There are still others who may return those with a core charge and resell others to core scavengers or scrap metal dealers.

Task F.8 Handle warranty and new returns.

Additional forms must usually be completed for warranty-return parts. This allows the manufacturers to determine what is wrong with the part and to credit the store for the return. A warranty-return part should never be placed back in the inventory to be sold to another customer.

Task F.9 Determine proper selling unit (each, pair, case, etc.) increment.

The term *each* refers to a single item. *Carton* or *case* refers to a number of parts packaged together and sold as a unit from the supplier. *Pair* refers to two of something. A *set* refers to two or more compatible items. A *kit* usually refers to a group of components used to rebuild a part.

Task F.10 Handle return of broken kits, special order parts, and exchange parts.

To maintain an adequate supply of good used parts for remanufacturing, a core charge may be added when buying a remanufactured part. This charge is refunded when the customer returns the used part. Freight charges may be added to emergency order parts to cover the cost of special transportation to the store. It is not uncommon to add a long-distance phone charge for special phone orders. Some parts may have a restocking fee that is charged to the customer when the parts are returned for a refund.

Task F.11 Account for store use items.

Keeping track of store use items is very important. Insuring that the store is prepared to make a sale includes the product on the shelf, a customer and a way to collect the money for the transaction. This seems fairly basic but it requires store employees to monitor the inventory in the store, insure that the right parts are in the right places and to insure that invoice paper is adequately stocked to provide sales records for the store and customer. Store Operations includes employee training that focuses on the ability to sell, stock, and replenish auto parts to many different customers. The requirement to track cores, deliveries, buy outs, and inventory are keys to success. The control of internal exchange, cleaning supplies, record keeping, and employee scheduling are all part of accounting for store use items.

G. Merchandising (2 Questions)

Task G.1 Understand display strategy.

The best place for a display is in an open area near the front of the store or department. This ensures that the display will catch the customer's eye as soon as he or she enters. The display should be large and noticeable and it must look appealing to the customer. This is usually accomplished by building the display out of the product being sold. Often the manufacturer of a product will send a display kit that can be assembled in the store and which may include banners, signs, and shelves. Displays are also often built at the end of an isle where the product is located so the customer will notice the product without entering the isle. Displays can be set up with the intent of the customer taking the actual item from the display. The display should be set up with care to keep it from falling apart as items are removed. A display also needs to be maintained for a neat appearance and to keep it looking appealing to the customer.

Task G.2 Display pricing.

A normal practice of display pricing is to display an item that may be specially priced for that week in an eye-catching display in an open part of the store or department, along with a noticeable sign that displays the special pricing. The product that is specially priced may still remain in its assigned shelf position but this additional display should let customers know this item is on special. A customer is more likely to notice a display with a posted price rather than notice an item on the shelf in its normal position. This kind of display and pricing is also used for the introduction of new items. A vendor may also choose to put up a display if overstocked in a certain item but choose not to lower the price. This will help promote the item and possibly lead to more sales. Display pricing can also be used to bring customers to the department or store by the distribution of flyers with an explanation and the price, by advertising on a billboard with the price, or signs in a window with the price.

Task G.3 Inspect and maintain shelf quantities and condition.

In order to keep shelves and items stocked and looking appealing to a customer the shelves must be well maintained. The shelves, as well as the items on the shelves, should not be dirty or dusty. Dusty items will not sell well because nobody wants to buy anything that they may feel is old and has been sitting around forever. The items on the shelves should be replenished on a regular basis to encourage repeat customers to count on you to have the items they need at all times. If shelves are messy and items are not easily found, or shelves are always empty from not restocking, often enough this will discourage a sale or a customer from revisiting your store or department again. The next item on a shelf should be pulled forward when the one in front of it is sold. This will give your shelf the appearance of being full and items will be easier to find for the next customer. When items are restocked the new items should be put behind the old ones. This will help keep dust from building up on items.

A shelf should be labeled to some extent to let people know which items go where. Displaying a price on a shelf where the item goes will help keep shelves in order because a customer will not have to take the item off the shelf to see how the price may differ from a similar item next to it. A customer may not always return the item to the space where it was removed, causing an unorganized appearance. Returned items should not be put back on the shelf if they are opened as this will give them a used appearance and make the customer uneasy about purchasing them. Items that are returned unopened can be returned to the shelf if they appear to be in good order.

Task G.4 Identify impulse, seasonal, and related items.

A seasonal item is something that a parts department may not normally stock during certain times of the year. Some seasonal items are stocked throughout the entire year but not in large quantities. An example of one of these items is windshield washer fluid, which is stocked more heavily in the winter season than the summer season but is still sold throughout the year. Another such example is car batteries, which are stocked all year even though battery failures occur more often during the colder parts of the year. Some other items used year-round but stocked more heavily in the winter include antifreeze, winter windshield wipers, and gasoline additives to prevent gas line freeze. Some items more heavily stocked in the summer include air conditioning refrigerant, windshield sun shades, and wax or polish. An impulse item is a product that the customer buys on the spur of the moment to fill a "want" rather than a "need" for the item. Some examples of impulse items include vehicle appearance-enhancing items such as pinstriping kits, chrome accessories, high-flow air filters to improve vehicle performance, and interior items like extra cup holders or seat covers.

Task G.5 Utilize sales aides.

Most merchandisers agree that the best location for impulse items is at a point between the customer's chest and eye level. By careful planning, the placement of products at appealing locations throughout the store greatly increases sales. The average customer is very sensitive to the prices of commonly purchased merchandise. While some products, such as motor oil, may be a high volume item for the merchant, they are seldom high profit items. Changing motor oil is the most popular task performed by the do-it-yourselfer (DIYer). Also, it is the reason most vehicles visit a garage. Motor oil is a product that is often sold at near cost as a price leader to attract customers.

5. Sample Test for Practice

Sample Test

Please note the letter and number in parentheses following each question. They match the overview in section 4 that discusses the relevant subject matter. You may want to refer to the overview using this cross-referencing key to help with questions posing problems for you.

1. Parts Specialist A says a 10 percent discount for $25.00 is $2.50. Parts Specialist B says a part bought for $10.00 and sold for $15.00 will generate a 33 percent profit. Who is right?
   A. A only
   B. B only
   C. Both A and B
   D. Neither A nor B (A.1)

2. If a customer returns a $100.00 part and is charged a 5 percent restocking fee, how much money is returned to the customer?
   A. $5.00
   B. $95.00
   C. $100.00
   D. $105.00 (A.2)

3. Parts Specialist A says that 350 cubic inches is about 5.0 liters. Parts Specialist B says that 390 cubic inches is about 6.6 liters. Who is right?
   A. A only
   B. B only
   C. Both A and B
   D. Neither A nor B (A.3)

4. Which of these numbers would appear first in an alphanumeric listing?
   A. 369482A
   B. 378654C
   C. 369482B
   D. 388426A (A.4)

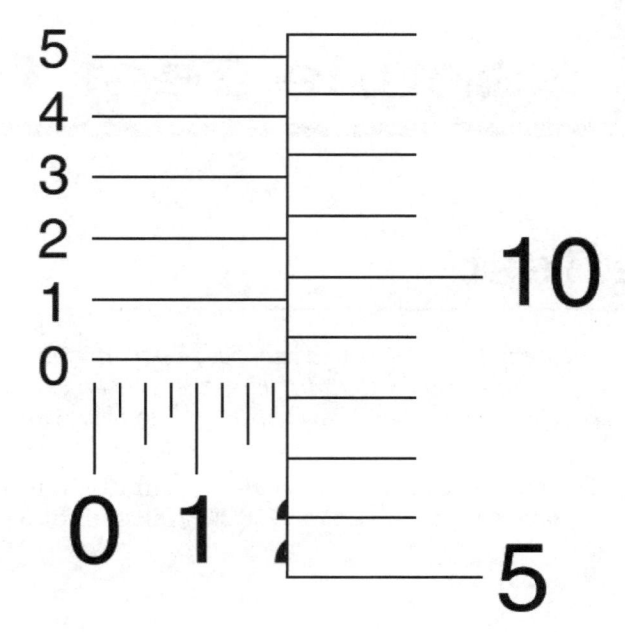

5. The 0-1 micrometer reading in the illustration is:
   A. 0.108 inch.
   B. 0.184 inch.
   C. 0.255 inch.
   D. 0.288 inch. (A.5)
6. Parts Specialist A says that charge accounts are commonly used for and encourage large sales. Parts Specialist B says that you must be familiar with the charge policies before you make the transaction. Who is right?
   A. A only
   B. B only
   C. Both A and B
   D. Neither A nor B (A.7)
7. Parts Specialist A says the ability to communicate is essential. Parts Specialist B says that interacting with all people is easy. Who is right?
   A. A only
   B. B only
   C. Both A and B
   D. Neither A nor B (A.8)
8. Parts Specialist A says a counterperson has the responsibility for keeping the entire store looking professional. Parts Specialist B says the store should have someone hired to clean up the shop. Who is right?
   A. A only
   B. B only
   C. Both A and B
   D. Neither A nor B (A.9)
9. Parts Specialist A says it is better to ask for help than to sell the customer the wrong parts. Parts Specialist B says that an experienced employee may be asked to help train a new employee. Who is right?
   A. A only
   B. B only
   C. Both A and B
   D. Neither A nor B (A.10)

PORT 2		PARTS-J3			INVOICING				10OCT00	1337

Invoice : 12223 Sale Type: CASH
Customer: 99999
Name: CASH RETAIL Parts: 158.40
Address: Freight: 0.00
City,St Zip: Tax: 9.50
Home Phone: Total Invoice: 167.90
 Backorder Amount: 0.00

Emp: 117   Sales Person:-   Ship Via: carry   PO: none   B/L:-

Part No.	Description	Bin	O.H.	Cost	Sale	Ext. sale	Q.S.	# A	O.O.	PM
23196550	WIPER MOTOR	245	11	62.36	87.30	87.30	1			
74073158	WIPER INSERTS	108	3	5.82	8.15	16.30	2			
21646146	WIPER SWITCH	218	8	37.38	52.33	52.33	1			
7148887	GROMMETS	13B	5	0.31	0.43	1.74	4			M
27289127	25 A FUSE	2A1	1	0.52	0.73	0.73	1			

F1=Help   F3=Save   F4=Cancel   F8=Print   F10=login

10. According to the invoice, what would be the total amount returned to the customer if the wiper motor was returned? Your shop has a policy to accept all returns providing they were not installed, are in their original package, and are not damaged.
    A. $87.30
    B. $62.36
    C. $92.54
    D. $66.10   (A.6)

11. A parts specialist should always:
    A. get help before heavy lifting.
    B. tell someone else to move heavy objects.
    C. size up the task before lifting an object.
    D. use a hand truck when lifting an object.   (A.11)

12. The Environmental Protection Agency (EPA) and the Occupational Safety and Health Administration (OSHA) give strict guidelines on all of the following chemicals **EXCEPT:**
    A. solvents.
    B. floor soaps.
    C. cutting oils.
    D. caustic cleaning compounds.   (A.12)

13. Parts Specialist A says that large items should be displayed on the top shelf so the customer can see them. Parts Specialist B says expensive items should be displayed behind the counter in locked cases. Who is right?
    A. A only
    B. B only
    C. Both A and B
    D. Neither A nor B   (A.13)

14. A do-it-yourselfer (DIYer) usually needs:
    A. a professional technician's assistance.
    B. used or rebuilt parts and accessories.
    C. rebuilt engines and transmissions.
    D. advice on parts and methods.   (B.1)

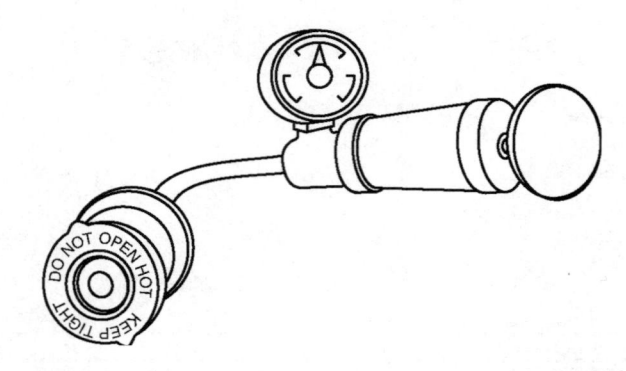

15. Parts Specialist A says the setup is used to check a radiator cap's ability to remain sealed as temperature changes. Parts Specialist B says the tester is checking the cap's ability to remain sealed as pressure increases. Who is right?
    A. A only
    B. B only
    C. Both A and B
    D. Neither A nor B   (C.2.4)

16. Parts Specialist A says that "What do you need?" is the best opening question to ask a customer. Parts Specialist B says that "May I help you?" is the best opening question to ask a customer. Who is right?
    A. A only
    B. B only
    C. Both A and B
    D. Neither A nor B   (B.2)

17. Parts Specialist A says that you should always establish who is to blame for a problem. Parts Specialist B says that you should raise your voice to stay above the angry customer's voice. Who is right?
    A. A only
    B. B only
    C. Both A and B
    D. Neither A nor B   (B.4)

18. Parts Specialist A says that the parts specialist should wait until it is the customer's turn before establishing eye contact. Parts Specialist B says that you should never talk to the customer while you are looking in the parts books. Who is right?
    A. A only
    B. B only
    C. Both A and B
    D. Neither A nor B   (B.5)

19. The phone rings while the parts specialist is helping a customer at the counter and there is no one else available to answer the phone. What should the parts specialist do? Parts Specialist A says to politely ask the counter customer to wait while you answer the phone. Parts Specialist B says to yell for assistance from another parts specialist. Who is right?
    A. A only
    B. B only
    C. Both A and B
    D. Neither A nor B   (B.6)

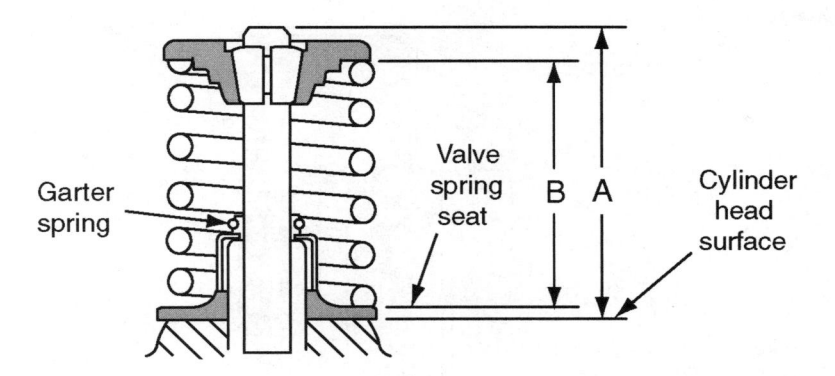

20. Which of the following best describes what dimension A in the drawing represents?
    A. Valve spring free height
    B. Valve spring installed height
    C. Valve stem height
    D. Valve retainer to seat clearance (C.1.4)

21. Parts Specialist A says that the appearance of the store as well as the appearance of the counter personnel influence the customer. Parts Specialist B says that a clean, neat parts specialist can influence the customer. Who is right?
    A. A only
    B. B only
    C. Both A and B
    D. Neither A nor B (B.8)

22. Parts Specialist A says that selling related parts can boost profits by as much as 30 percent. Parts Specialist B says that you should always promote stock that is not selling well. Who is right?
    A. A only
    B. B only
    C. Both A and B
    D. Neither A nor B (B.9)

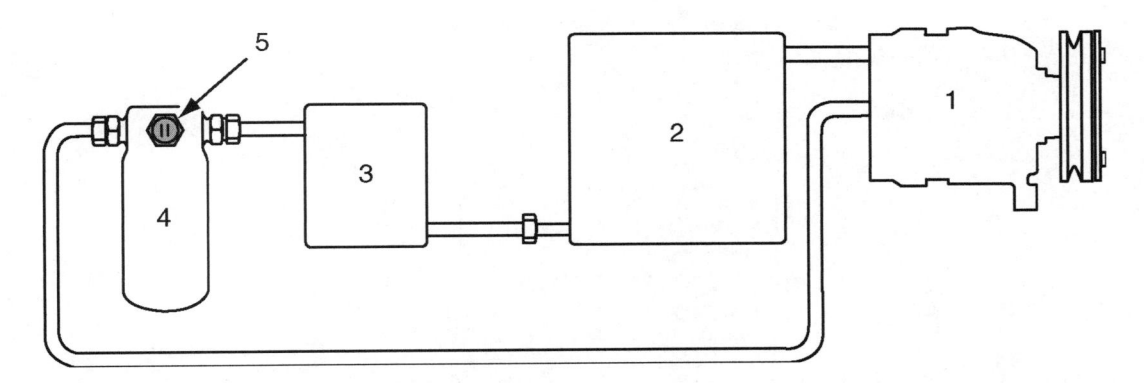

23. In the figure, Parts Specialist A says the component labeled 3 is the evaporator. Parts Specialist B says the component labeled 1 is the compressor. Who is right?
    A. A only
    B. B only
    C. Both A and B
    D. Neither A nor B (C.12.1)

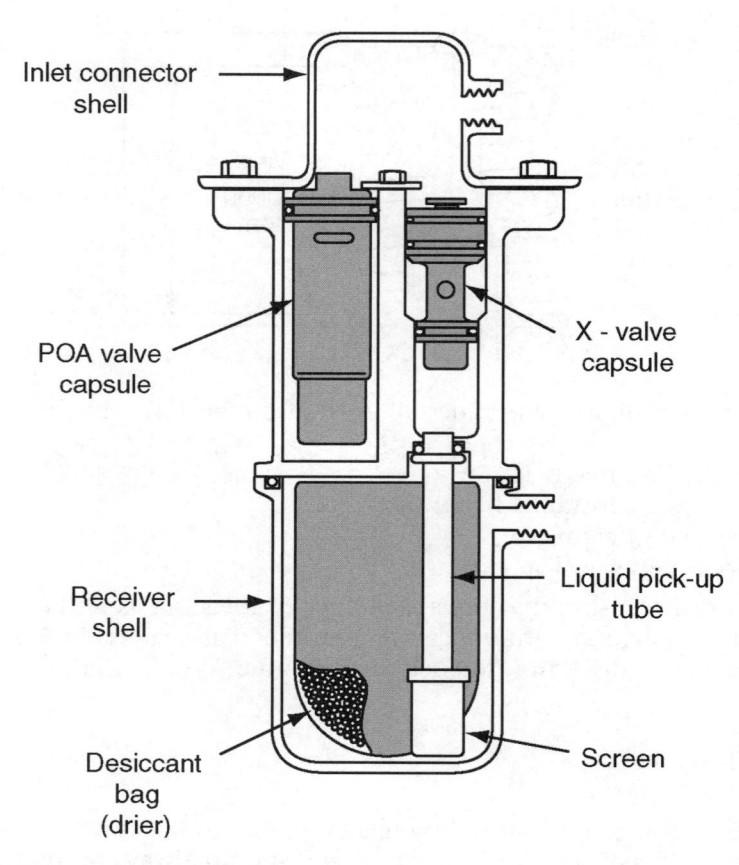

[Figure: Component with labels — Inlet connector shell, POA valve capsule, X-valve capsule, Receiver shell, Liquid pick-up tube, Desiccant bag (drier), Screen]

24. The component in the figure is a(n):
    A. accumulator.
    B. receiver/drier.
    C. VIR (valves in receiver) unit.
    D. refrigerant filter and valve assembly. (C.12.1)

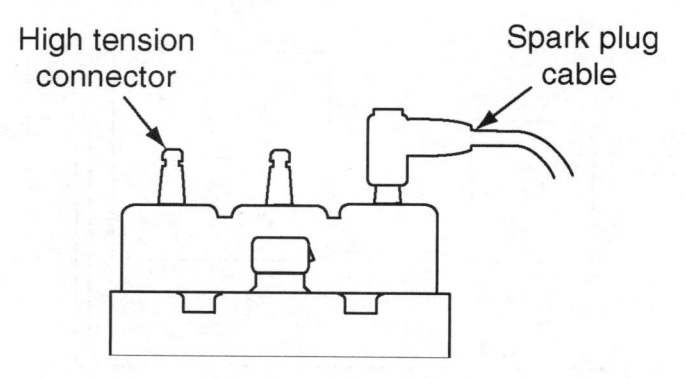

[Figure: Ignition coil pack with High tension connector and Spark plug cable labeled]

25. Parts Specialist A says the part shown in the figure is the ignition coil pack for at least three cylinders. Parts Specialist B says the part shown is an ignition coil pack for a distributorless ignition system. Who is right?
    A. A only
    B. B only
    C. Both A and B
    D. Neither A nor B (C.4.1)

26. All of the following components are in the fuel system **EXCEPT:**
    A. the fuel pump.
    B. the fuel tank.
    C. the carburetor.
    D. fuel tank hangers. (C.3.1)

27. Parts Specialist A says that an EGR valve cools the combustion temperatures during light engine loads. Parts Specialist B says that EGR valves are used for the cruise control system. Who is right?
    A. A only
    B. B only
    C. Both A and B
    D. Neither A nor B (C.6.2)

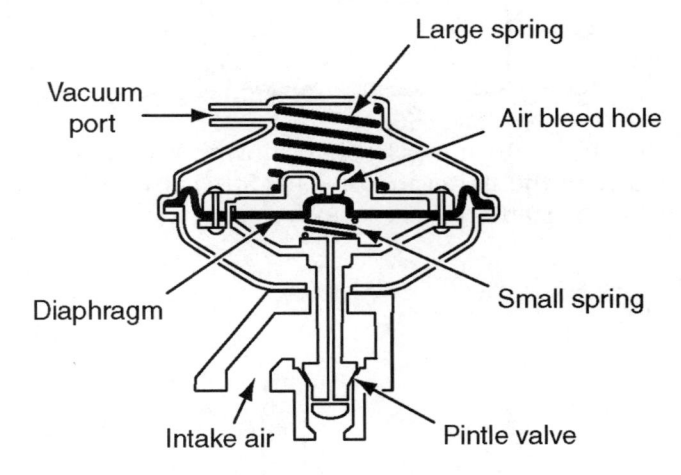

28. Parts Specialist A says the component shown in the figure is a major component of the emission control system. Parts Specialist B says the part is an EGR valve. Who is right?
    A. A only
    B. B only
    C. Both A and B
    D. Neither A nor B (C.6.1)

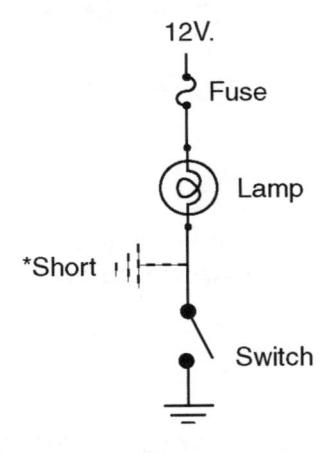

29. Which of the following statements is true about the short shown in the drawing?
    A. The fuse would blow.
    B. The lightbulb would burn out quickly.
    C. The lamp would remain on at all times.
    D. The switch would heat up due to arcing across the contacts. (C.13.4)

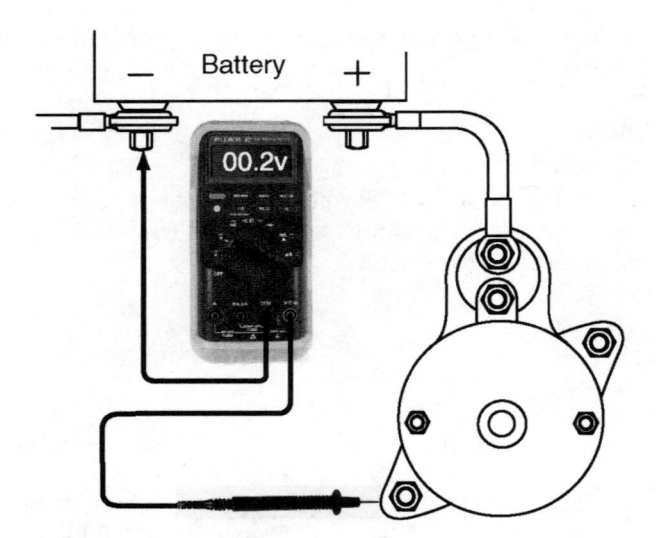

30. Parts Specialist A says the meter hookup shown in the drawing is measuring the available voltage to the starter motor. Parts Specialist B says the reading of 0.2 volts shows the starter is bad. Who is right?
    A. A only
    B. B only
    C. Both A and B
    D. Neither A nor B

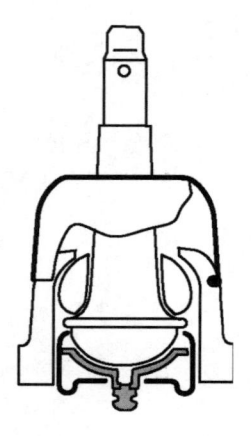

(C.13.4)

31. The suspension part shown in the figure is a(n):
    A. tie-rod end.
    B. ball joint.
    C. idler arm.
    D. wheel spindle.

(C.11.1)

32. Parts Specialist A says that the suspension keeps the vehicle in control of the road. Parts Specialist B says that the suspension helps to cushion the vehicle from road shock. Who is right?
    A. A only
    B. B only
    C. Both A and B
    D. Neither A nor B

(C.11.2)

Camber

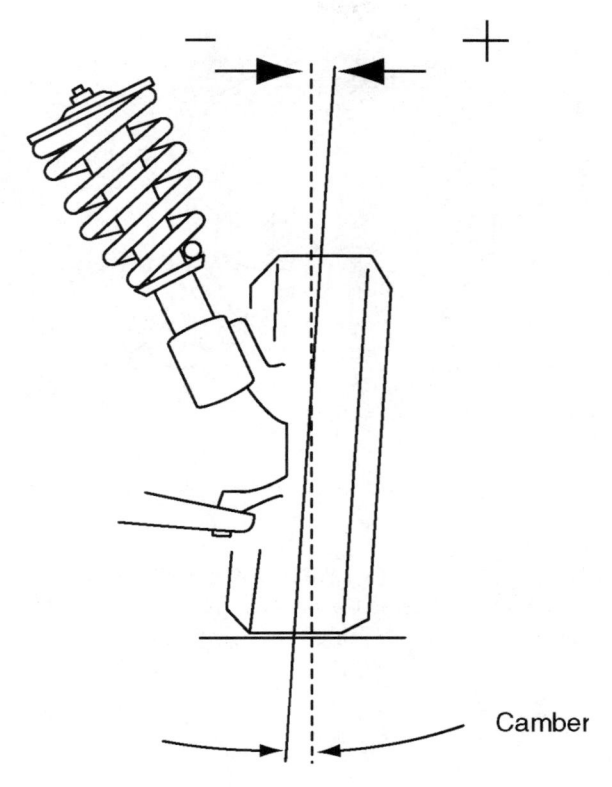

33. Referring to the drawing, Parts Specialist A says a bent strut will cause this alignment angle to be out of specs. Parts Specialist B says if this angle is wrong, excessive tire wear will result. Who is right?
   A. A only
   B. B only
   C. Both A and B
   D. Neither A nor B   (C.11.4)

34. Parts Specialist A says that the starter motor turns a flywheel mounted on the rear of the crankshaft. Parts Specialist B says that when the ignition switch is turned to START, there is electrical power directly to the alternator. Who is right?
   A. A only
   B. B only
   C. Both A and B
   D. Neither A nor B   (C.13.2)

35. Bolt dimensions are determined by the:
   A. distance across the flats of the hex head.
   B. size wrench that fits on the bolt head.
   C. distance across the points of the hex head.
   D. diameter and pitch of the bolt threads.   (C.14.3)

36. Parts Specialist A says that the location of the vehicle identification number (VIN) tag is on the driver's side inner fender. Parts Specialist B says that the vehicle identification number (VIN) tag is visible through the driver's door window. Who is right?
   A. A only
   B. B only
   C. Both A and B
   D. Neither A nor B   (D.1)

Wooden dowel

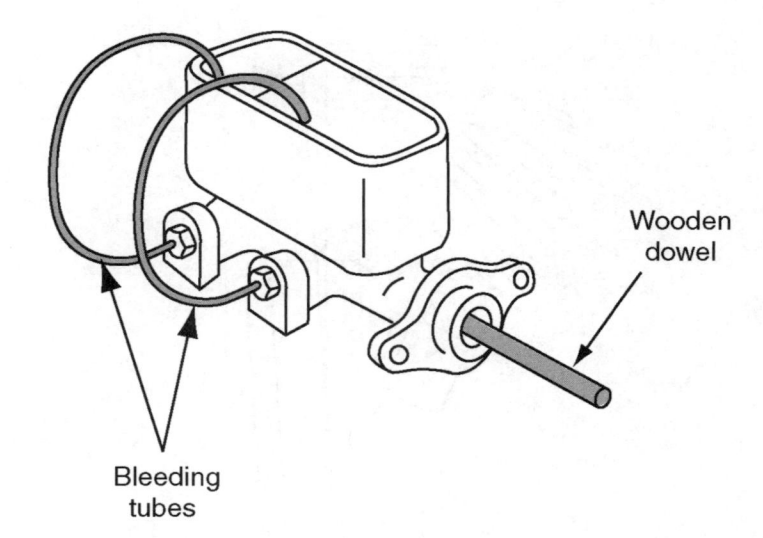

Bleeding tubes

37. The setup shown in the drawing, Parts Specialist A says the procedure is called bench bleeding. Parts Specialist B says the procedure is bench bleeding and is done to eliminate the need to bleed the entire brake system after a master cylinder is installed in a vehicle. Who is right?
    A. A only
    B. B only
    C. Both A and B
    D. Neither A nor B (C.10.2)

38. Parts Specialist A says that the paint code is in the vehicle identification number (VIN). Parts Specialist B says that the paint code can be found by looking in the owner's manual. Who is right?
    A. A only
    B. B only
    C. Both A and B
    D. Neither A nor B (D.6)

39. Parts Specialist A says that each catalog has a form number. Parts Specialist B says that the cover to each catalog should identify the product and manufacturer. Who is right?
    A. A only
    B. B only
    C. Both A and B
    D. Neither A nor B (E.1)

40. Parts Specialist A says that footnotes can be very important for getting the correct part. Parts Specialist B says that footnotes are provided only for special order parts. Who is right?
    A. A only
    B. B only
    C. Both A and B
    D. Neither A nor B (E.2)

41. Three types of fittings used on automobile hoses and tubes are:
    A. flare, compression, and tubing.
    B. pipe, flare, and steel.
    C. tubing, steel, and pipe.
    D. flare, compression, and pipe. (C.14.5)

42. All of the following are types of bulletins **EXCEPT**:
    A. new item availability bulletins.
    B. delivery bulletins.
    C. product information bulletins.
    D. correction bulletins. (E.3)

ABS	- Anti-lock Brake System	FWD	- Front Wheel Drive	P	- Passenger's Side
AC	- Air Conditioning	I.D.	- Inside Diameter	Qty.	- Quantity
AT	- Automatic Transmission	Man	- Manual	RWD	- Rear Wheel Drive
ATC	- Automatic Temperature Control	min.	- Minimum	Veh.	- Vehicle
		mm	- Millimeters	w/	- With
Auto	- Automatic	Max.	- Maximum	w/o	- Without
cid	- Cubic Inch Displacement	MT	- Manual Transmission	3AUT	- 3 Speed Automatic Transmission
cc	- Cubic Centimeters	NA	- Not Available	4AUT	- 4 Speed Automatic Transmission
Comp.	- Competitor	No.	- Number	4MAN	- 4 Speed Manual Transmission
C.V.	- Constant Velocity	NR	- Not Required	4WD	- Four Wheel Drive
D	- Driver's Side	O.D.	- Outside Diameter	5MAN	- 5 Speed Manual Transmission
dia.	- Diameter	O.E.	- Original Equipment	6MAN	- 6 Speed Manual Transmission
exc	- Except				

43. In the figure, the abbreviation AC means:
    A. automatic controls.
    B. asbestos contamination.
    C. air conditioning.
    D. air compressor. (E.4)

44. Parts Specialist A says that there is no reason to rotate stock on items that are not perishable. Parts Specialist B says that if the price is displayed on a shelf it may help to maintain the appearance of the store or department. Who is right?
    A. A only
    B. B only
    C. Both A and B
    D. Neither A nor B (G.3)

45. All of the following are true about display pricing **EXCEPT**:
    A. Display pricing is often used for the introduction of new items.
    B. Display pricing is often used for special sale items.
    C. The customer is more likely to notice the price and product in a display rather than on a shelf.
    D. If an item is in a display then it will always be on sale. (G.2)

46. A display arrangement must be:
    A. no higher than four feet.
    B. in an open area to be effective.
    C. near where the product usually is.
    D. set up by a display person. (G.1)

47. A customer asks for four liters of oil. How many quarts should the parts specialist give the customer?
    A. Two
    B. Four
    C. Five
    D. Six (A.3)

48. Parts Specialist A says that an invoice should be completed for each sale. Parts Specialist B says that back orders are the type of orders made to maintain the present stock in the store. Who is right?
    A. A only
    B. B only
    C. Both A and B
    D. Neither A nor B (A.14)

49. Parts Specialist A says that some stores have policies that will not allow the customer to return parts that have been installed. Parts Specialist B says that most do-it-yourselfers have good enough diagnostic skills to replace the correct part. Who is right?
   A. A Only
   B. B Only
   C. Both A and B
   D. Neither A nor B (B.7)

50. A customer asks for a set of brake shoes and the parts specialist has four different kinds available. What should the parts specialist do next?
   A. Sell the customer the best set of brake shoes.
   B. Sell the customer the most expensive set of brake shoes.
   C. Tell the customer about the different kinds of brake shoes and let the customer make the decision.
   D. Sell the customer the cheapest set of brake shoes because the vehicle is old and not worth very much. (B.10)

51. Parts Specialist A says that a parts specialist should never push a sale. Parts Specialist B says that it is part of a parts specialist's job to identify whether the customer is there to look around or there to buy something. Who is right?
   A. A only
   B. B only
   C. Both A and B
   D. Neither A nor B (B.11)

52. Parts Specialist A says that if you are helping a customer at the counter and the telephone rings, you should just push the hold button. Parts Specialist B says that you should never interrupt activities with a customer at the counter by answering the telephone. Who is right?
   A. A only
   B. B only
   C. Both A and B
   D. Neither A nor B (B.12)

53. A customer asks for brake shoes. While looking up the right application for the brake shoes the parts specialist should:
   A. not take the time to answer the phone.
   B. finish lunch (if during lunchtime).
   C. talk to other counter personnel.
   D. suggest related items for the brake repair. (B.13)

54. Parts Specialist A says that if a DIY customer asks for brake shoes it is not necessary to ask if brake fluid is needed. Parts Specialist B says that suggesting related items is only done when talking to a trained repair technician. Who is right?
   A. A only
   B. B only
   C. Both A and B
   D. Neither A nor B (B.14)

55. A customer comes into the store carrying a set of jumper cables. The customer asks for a starter motor for his late model vehicle. What should the parts specialist do next?
   A. Look up the starter motor for the vehicle.
   B. Ask the customer what symptoms suggested the need for a starter.
   C. Refer the customer to a good automotive repair facility.
   D. Ask another parts specialist what starters fit the customer's vehicle. (B.15)

56. Closing the sale means:
   A. getting a commitment from the customer.
   B. taking the customer's money.
   C. looking up the correct part.
   D. handing the customer their change and receipt. (B.16)

57. Parts Specialist A says that the vehicle build sheet can indicate the vehicle color. Parts Specialist B says that the vehicle build sheet can indicate the vehicle engine size. Who is right?
   A. A only
   B. B only
   C. Both A and B
   D. Neither A nor B (D.5)

58. Parts Specialist A says that the table of contents will show the catalog sequence. Parts Specialist B says that the index will show the catalog sequence. Who is right?
   A. A only
   B. B only
   C. Both A and B
   D. Neither A nor B (E.5)

59. Parts Specialist A says that bulletins are the primary tool for conducting catalog maintenance. Parts Specialist B says that bulletins are not very useful and that posting them is a waste of effort. Who is right?
   A. A only
   B. B only
   C. Both A and B
   D. Neither A nor B (E.6)

60. Parts Specialist A says that a packing list should be included with all deliveries. Parts Specialist B says that the packing list gives directions for where to deliver the package. Who is right?
   A. A only
   B. B only
   C. Both A and B
   D. Neither A nor B (F.2)

61. Physical inventory should be done:
   A. every day.
   B. weekly.
   C. monthly.
   D. annually. (F.3)

62. Physical inventory discrepancies are reported by:
   A. telling the store manager.
   B. calling the jobber and letting them know about the discrepancy.
   C. filling out an inventory discrepancy form.
   D. letting the other parts specialists know about the discrepancies. (F.4)

63. Parts Specialist A says that stock rotations are not necessary. Parts Specialist B says that stock rotation may be done annually. Who is right?
   A. A only
   B. B only
   C. Both A and B
   D. Neither A nor B (F.5)

64. A special order is:
   A. the order that is made every week.
   B. a part ordered because the store does not keep it in stock.
   C. an order made because a vehicle is inoperative and needs to be returned to service as soon as possible.
   D. an order placed for the parts manager or store owner. (F.6)

65. A core charge is:
    A. a different name for a state sales tax.
    B. the charge added by the store above the usual markup.
    C. a charge added when a customer buys a remanufactured part.
    D. the charge for restocking a returned part. (F.7)

66. Parts Specialist A says that an additional form must be filled out for warranty returns. Parts Specialist B says that a part returned for warranty should not be put back on the shelf and sold to another customer. Who is right?
    A. A only
    B. B only
    C. Both A and B
    D. Neither A nor B (F.8)

67. A case refers to:
    A. a package of five parts.
    B. the number of parts packaged together and sold as a unit.
    C. the number of parts in a display box.
    D. a package of twenty pieces. (F.9)

68. A restocking fee may be charged to a customer for:
    A. buying too many of one type of a part.
    B. having the brake rotors turned.
    C. not buying enough parts each month.
    D. returning a part for a refund. (F.10)

69. Parts Specialist A says that there are no benefits to using lost sales reports. Parts Specialist B says that most stores analyze not only the parts that they are selling, but parts that the store does not sell so well. Who is right?
    A. A only
    B. B only
    C. Both A and B
    D. Neither A nor B (F.1)

70. Parts Specialist A says excessive flywheel run out may cause grabbing or erratic clutch operation. Parts Specialist B says the pressure plate should always be reinstalled in the original position on the flywheel. Who is right?
    A. A only
    B. B only
    C. Both A and B
    D. Neither A nor B (C.1.3)

71. The following are normal oil pump component measurements **EXCEPT:**
    A. inner rotor diameter.
    B. clearance between the rotors.
    C. inner and outer rotor thickness.
    D. outer rotor to housing clearance. (C.2.3)

72. Parts Specialist A says a defective water pump bearing may cause a growling noise when the engine is idling. Parts Specialist B says the water pump bearing may be ruined by coolant leaking past the pump seal. Who is right?
    A. A only
    B. B only
    C. Both A and B
    D. Neither A nor B (C.2.4)

73. Parts Specialist A says an intake manifold vacuum leak may cause a cylinder misfire with the engine idling. Parts Specialist B says an intake manifold vacuum leak may cause a cylinder misfire during hard acceleration. Who is right?
    A. A only
    B. B only
    C. Both A and B
    D. Neither A nor B   (C.3.3)

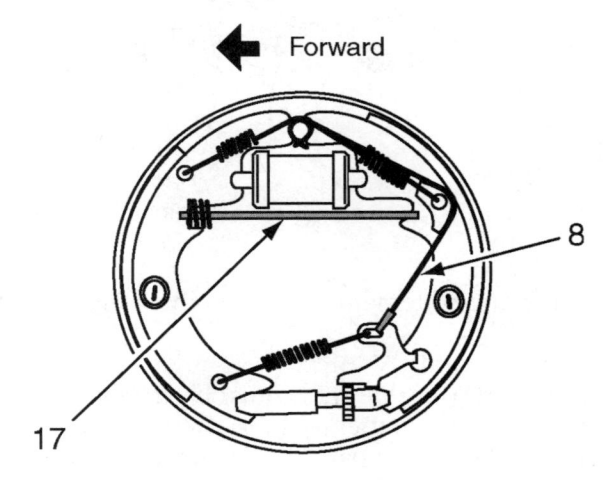

74. Parts Specialist A says the part labeled 8 in the drawing is part of the self-adjusting function of the brake shoe assembly. Parts Specialist B says the part labeled 17 keeps the two brake shoes aligned while they expand and retract. Who is right?
    A. A only
    B. B only
    C. Both A and B
    D. Neither A nor B   (C.10.2)

75. The following statements about distributor advances are true **EXCEPT:**
    A. The vacuum advance controls spark advance in relation to engine load.
    B. The mechanical advance controls spark advance in relation to engine rpm.
    C. The mechanical advance rotates the reluctor in the opposite direction to shaft rotation.
    D. The vacuum advance rotates the pickup plate in the opposite direction to shaft rotation.   (C.4.3)

76. All of the following statements regarding manifold heat control valves are true **EXCEPT:**
    A. A manifold heat control valve improves fuel vaporization in the intake manifold especially when the engine is cold.
    B. A manifold heat control valve stuck in the closed position causes a loss of engine power.
    C. A manifold heat control valve stuck in the open position may cause an acceleration stumble.
    D. A manifold heat control valve stuck in the closed position reduces intake manifold temperature.   (C.5.1)

Snap ring pliers

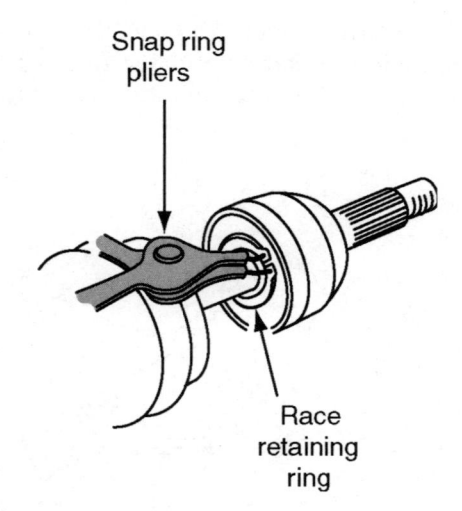

Race retaining ring

77. Which of the following statements about the part shown in the figure is NOT true?
   A. The part is an outboard CV joint.
   B. To remove the part, the axle shaft must be removed first.
   C. The part is typically permanently lubricated.
   D. A boot is installed over the part to prevent moisture and dirt from damaging it. (C.9.4)

78. The shift lever adjustment usually is performed with the transmission in:
   A. neutral.
   B. first gear.
   C. second gear.
   D. reverse gear. (C.7.3)

79. A five-speed manual transaxle has a growling and rattling noise in third gear only. The cause of this noise could be worn:
   A. or chipped teeth on the third speed gear on the input shaft.
   B. or chipped dog teeth on the third gear on the input shaft.
   C. dog teeth on the third speed synchronizer blocking ring.
   D. threads in the cone area of the third speed blocking ring. (C.7.4)

80. An automatic transmission has a whining noise that occurs in all gears while the vehicle is being driven. This noise is also present with the engine running and the vehicle stopped. Parts Specialist A says the rear planetary gearset may be defective. Parts Specialist B says the oil pump may be defective. Who is right?
   A. A only
   B. B only
   C. Both A and B
   D. Neither A nor B (C.8.3)

81. Parts Specialist A says that improper shift linkage adjustments may cause premature transmission clutch failure. Parts Specialist B says that improper shift linkage adjustments may cause higher than normal fluid pressure. Who is right?
   A. A only
   B. B only
   C. Both A and B
   D. Neither A nor B (C.8.4)

82. A front-wheel-drive car has a clunking noise while decelerating. Parts Specialist A says this noise may be caused by a worn inner drive axle joint. Parts Specialist B says this noise may be caused by a worn front wheel bearing. Who is right?
   A. A only
   B. B only
   C. Both A and B
   D. Neither A nor B   (C.9.3)

83. Parts Specialist A says a damaged brake line may be repaired with a short piece of line and compression fittings. Parts Specialist B says the necessary brake line bends should be made with a tube-bending tool. Who is right?
   A. A only
   B. B only
   C. Both A and B
   D. Neither A nor B   (C.10.3)

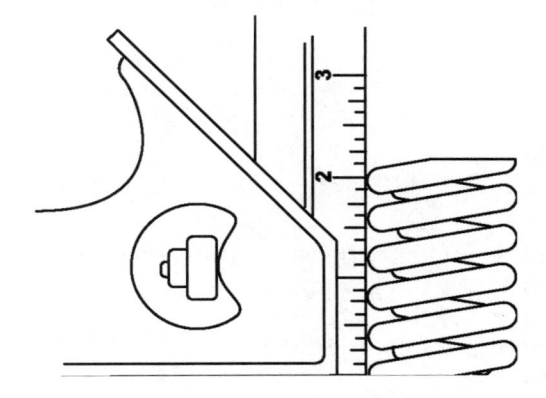

84. While discussing the setup shown in the figure, Parts Specialist A says the freestanding height of the valve spring is being measured. Parts Specialist B says the spring is being checked for squareness. Who is right?
   A. A only
   B. B only
   C. Both A and B
   D. Neither A nor B   (C.1.3)

85. When one side of the front or rear bumper is pushed downward with considerable weight and then released, the bumper makes two free upward bounces before the vertical chassis movement stops. This action indicates:
   A. a defective shock absorber.
   B. a weak coil spring.
   C. a broken spring insulator.
   D. a worn stabilizer bushing.   (C.11.3)

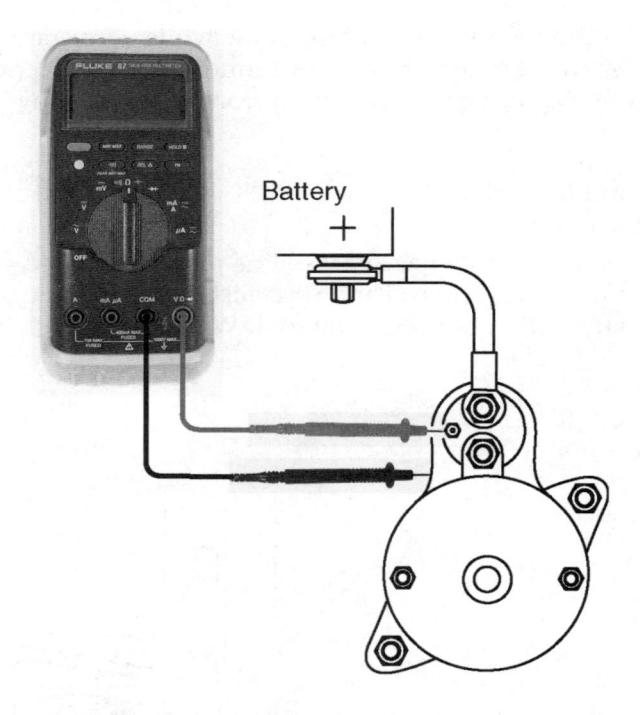

86. Which of the following describes what the meter setup in the figure is measuring?
    A. Voltage drop across the ground circuit
    B. Available voltage at the starter
    C. Voltage drop of the solenoid
    D. Voltage at the battery (C.13.2)

87. Parts Specialist A says restricted refrigerant passages in the evaporator may cause frosting of the evaporator outlet pipe. Parts Specialist B says restricted refrigerant passages in the evaporator may cause much higher than specified low-side pressures. Who is right?
    A. A only
    B. B only
    C. Both A and B
    D. Neither A nor B (C.12.3)

88. When testing a diode, connect the ohmmeter leads across the diode and then reverse the leads. A satisfactory diode provides:
    A. two high meter readings.
    B. one high one low meter reading.
    C. two low meter readings.
    D. a meter reading of 0 Ohms and 10 Ohms. (C.13.4)

89. Parts Specialist A says that a good location for merchandise to attract impulse sales is on a shelf level that is between the knees and waist of the customer. Parts Specialist B says a good location for merchandise to attract impulse sales is on a shelf that is just above the customer's eye level. Who is right?
    A. A only
    B. B only
    C. Both A and B
    D. Neither A nor B (G.5)

90. Parts Specialist A says there is more than one coil in a distributorless ignition system (DIS). Parts Specialist B says the coil synchronizes the ignition module in relation to the crankshaft position. Who is right?
   A. A only
   B. B only
   C. Both A and B
   D. Neither A nor B   (C.4.2)

91. The following are all part of the exhaust system **EXCEPT:**
   A. the muffler.
   B. the resonator.
   C. the catalytic converter.
   D. the reverberator.   (C.5.2)

92. Parts Specialist A says that pressing the clutch pedal engages the clutch. Parts Specialist B says the clutch pressure plate is operated by a clutch release bearing. Who is right?
   A. A only
   B. B only
   C. Both A and B
   D. Neither A nor B   (C.7.2)

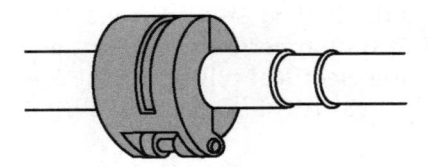

← Push tool into cage opening

93. Parts Specialist A says the tool in the figure is required to join two ends of a refrigerant line together. Parts Specialist B says this tool can be used to cut the line so a defective section could be removed. Who is right?
   A. A only
   B. B only
   C. Both A and B
   D. Neither A nor B   (C.12.3)

94. Parts Specialist A says that the refrigerant leaving a compressor is a high-pressure vapor. Parts Specialist B says that the refrigerant entering the compressor is a low pressure vapor. Who is right?
   A. A only
   B. B only
   C. Both A and B
   D. Neither A nor B   (C.12.2)

95. The following are all part of the accessory electrical system **EXCEPT:**
   A. a radio or sound system.
   B. an alternator or generator.
   C. power seats or windows.
   D. a rear window defogger.   (C.13.1)

96. Parts Specialist A says that all bolts securing a part should be of the same grade. Parts Specialist B says that a nut should be the same grade as the bolt it is used on. Who is right?
   A. A only
   B. B only
   C. Both A and B
   D. Neither A nor B   (C.14.4)

97. Which type fitting will include a ferrule?
    A. Compression
    B. Pipe
    C. Flare
    D. Flange (C.14.5)

98. Parts Specialist A says that all hoses used in automotive service must withstand high pressure. Parts Specialist B says that some hoses used in automotive service are made of a reinforced synthetic rubber. Who is right?
    A. A only
    B. B only
    C. Both A and B
    D. Neither A nor B (C.14.8)

99. Parts Specialist A says that the open end wrench will grip on two sides of the bolt only. Parts Specialist B says that the box end wrench will grip on two sides of the bolt only. Who is right?
    A. A only
    B. B only
    C. Both A and B
    D. Neither A nor B (A.16)

100. Parts Specialist A says the battery stores the electrical energy. Parts Specialist B says the electrical/charging system performs two duties, maintaining the battery's state of charge and providing electrical power for the ignition system, air conditioner, heater, lighting, and all electrical accessories. Who is right?
    A. A only
    B. B only
    C. Both A and B
    D. Neither A nor B (C.14.1)

101. Parts Specialist A says that the cold cranking amps is the number of volts the battery can deliver at zero degrees to start the engine. Parts Specialist B says that the reserve amperage is the output of amperage while there is not charging input. Who is right?
    A. A only
    B. B only
    C. Both A and B
    D. Neither A nor B (C.14.3)

102. Parts Specialist A sells a part with a core, but does not include the core charge on the invoice. She says that it is correctly billed and that the core will be returned. Parts Specialist B says that there is no way to track the returned core to determine whether it has been returned or not. Who is right?
    A. A only
    B. B only
    C. Both A and B
    D. Neither A nor B (F.11)

6 Additional Test Questions for Practice

Additional Test Questions

Please note the letter and number in parentheses following each question. They match the overview in section 4 that discusses the relevant subject matter. You may want to refer to the overview using this cross-referencing key to help with questions posing problems for you.

1. Parts Specialist A says a 20 percent discount on a $60.00 part is $12.00. Parts Specialist B says a part bought for $100.00 and sold for $175.00 will generate a 50 percent profit. Who is right?
   A. A only
   B. B only
   C. Both A and B
   D. Neither A nor B (A.1)

2. If a customer returns a $250.00 part and is charged a 10 percent restocking fee, how much money is returned to the customer?
   A. $50.00
   B. $100.00
   C. $1000.00
   D. $225.00 (A.2)

3. Parts Specialist A says a 5-liter engine has a displacement of 350 cubic inches. Parts Specialist B says a thermostat marked 192° F is using metric units. Who is right?
   A. A only
   B. B only
   C. Both A and B
   D. Neither A nor B (A.3)

4. Which of these numbers would appear first in an alphanumeric listing?
   A. 88972E
   B. 89972B
   C. 89973A
   D. 88973A (A.4)

91

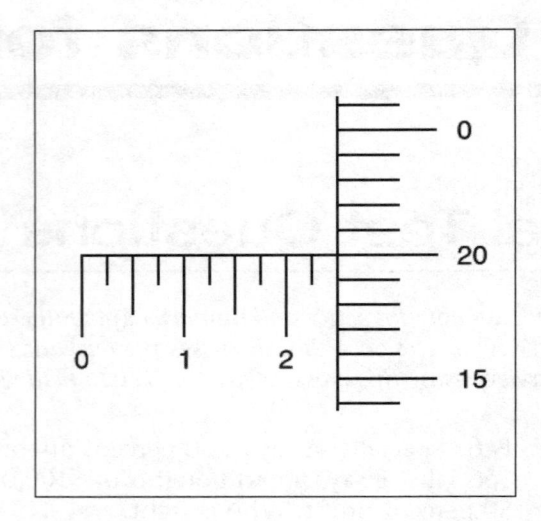

5. The 0–1 inch micrometer reading shown in the illustration is:
   A. 0.220 inch.
   B. 0.225 inch.
   C. 0.245 inch.
   D. 0.250 inch. (A.5)
6. Parts Specialist A says that the parts specialist is responsible for keeping the floors clean. Parts Specialist B says that the parts specialist is only responsible for helping customers. Who is right?
   A. A only
   B. B only
   C. Both A and B
   D. Neither A nor B (A.9)
7. Parts Specialist A says that any company with hazardous chemicals must maintain Material Safety Data Sheets (MSDSs). Parts Specialist B says that cutting oil is not a chemical that is hazardous. Who is right?
   A. A only
   B. B only
   C. Both A and B
   D. Neither A nor B (A.12)

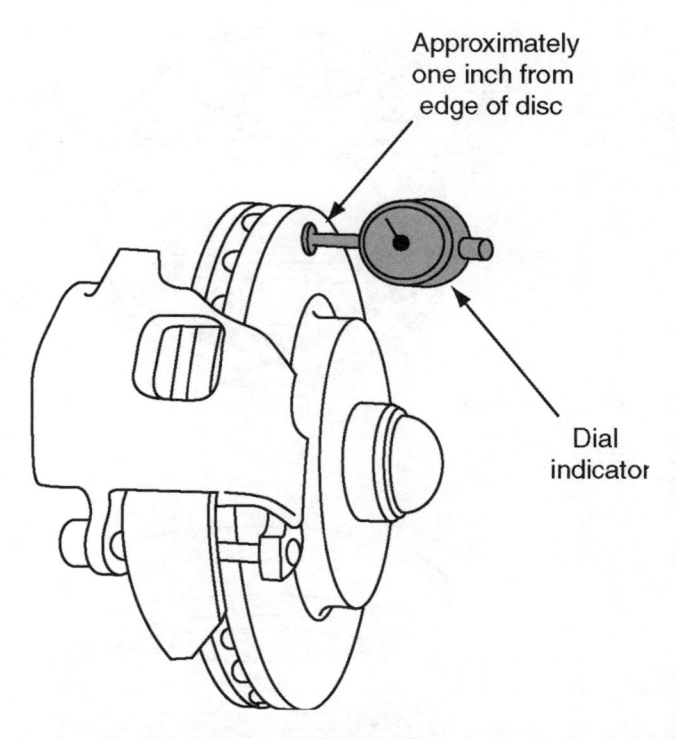

Additional Test Questions for Practice Additional Test Questions 93

Approximately one inch from edge of disc

Dial indicator

8. Parts Specialist A says the setup is checking the run out of a brake rotor. Parts Specialist B says the setup is measuring the rotor thickness. Who is right?
   A. A only
   B. B only
   C. Both A and B
   D. Neither A nor B   (C.10.4)

9. What kind of merchandise should be displayed by the exit of the store?
   A. Spark plugs
   B. Barrels of oil
   C. Lug nuts
   D. Valve stems   (A.13)

10. Parts Specialist A says that you can assist a do-it-yourselfer without taking too much time by having how-to information at hand. Parts Specialist B says that it is too time consuming to help customers who do not know what they need. Who is right?
    A. A only
    B. B only
    C. Both A and B
    D. Neither A nor B   (B.1)

11. Parts Specialist A says that an opening statement should find out what the customer needs. Parts Specialist B says that an opening statement should put the customer at ease. Who is right?
    A. A only
    B. B only
    C. Both A and B
    D. Neither A nor B   (B.2)

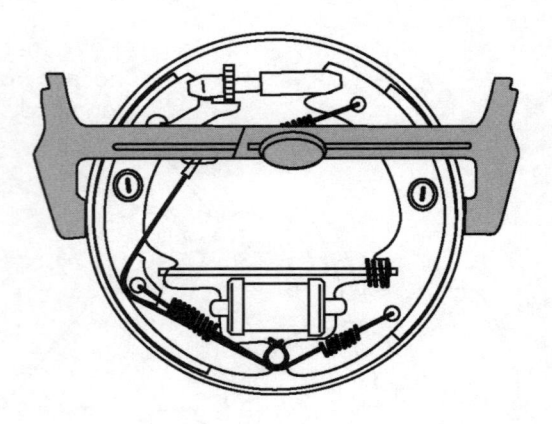

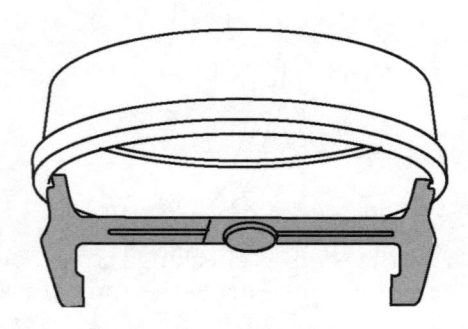

12. While discussing the procedure shown in the figure, Parts Specialist A says the wear of the brake shoes and brake drum is being checked. Parts Specialist B says the procedure is used to set the brake shoes prior to installation. Who is right?
    A. A only
    B. B only
    C. Both A and B
    D. Neither A nor B   (C.1.4)

13. Parts Specialist A says that a counter customer is more important than a telephone customer. Parts Specialist B say that a telephone customer is more important than putting the stock order away. Who is right?
    A. A only
    B. B only
    C. Both A and B
    D. Neither A nor B   (B.6)

14. Parts Specialist A says that the appearance of the store has little influence on the customer. Parts Specialist B says that it is not the parts specialist's responsibility to keep the counter clean. Who is right?
    A. A only
    B. B only
    C. Both A and B
    D. Neither A nor B   (B.8)

15. Parts Specialist A says that if you sell the customers what they need, they will return again and again. Parts Specialist B says that you should promote the sale of items that are on sale. Who is right?
    A. A only
    B. B only
    C. Both A and B
    D. Neither A nor B   (B.9)

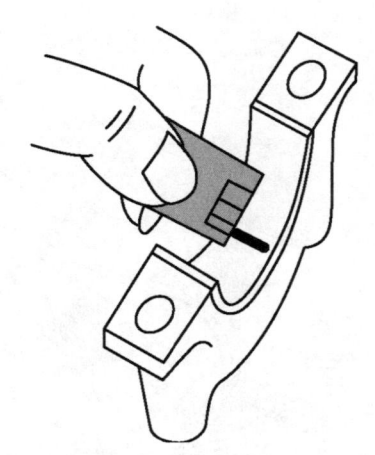

16. Parts Specialist A says the results of this procedure will determine if the bearing needs to be replaced. Parts Specialist B says this procedure will determine if the shaft that revolves on the bearing is out-of-round. Who is right?
    A. A only
    B. B only
    C. Both A and B
    D. Neither A nor B   (C.1.4)

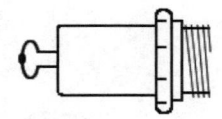

17. What part of the fuel system is shown in the figure?
    A. Fuel filter
    B. Fuel pump
    C. Fuel tank
    D. Fuel pressure regulator   (C.3.1)

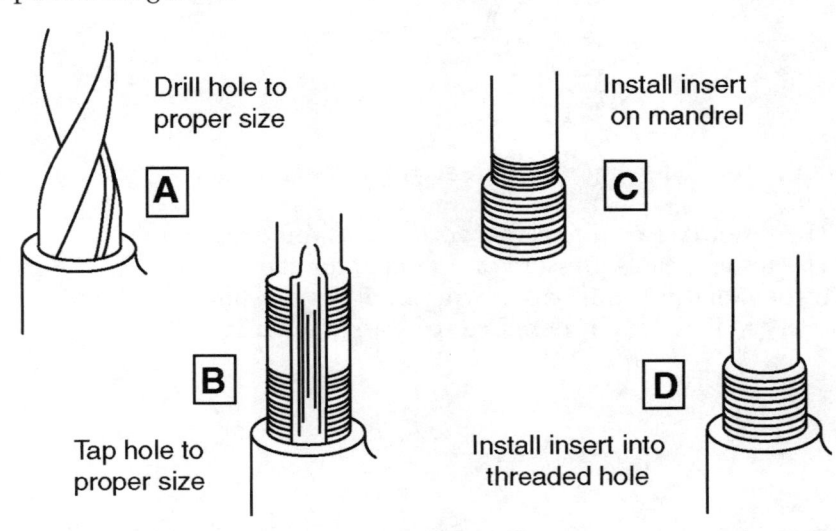

A. Drill hole to proper size
B. Tap hole to proper size
C. Install insert on mandrel
D. Install insert into threaded hole

18. Parts Specialist A says the procedure shown is used to repair damaged internal threads. Parts Specialist B says inserts like those being installed in the drawing are available in many different sizes. Who is right?
    A. A only
    B. B only
    C. Both A and B
    D. Neither A nor B   (C.14.2)

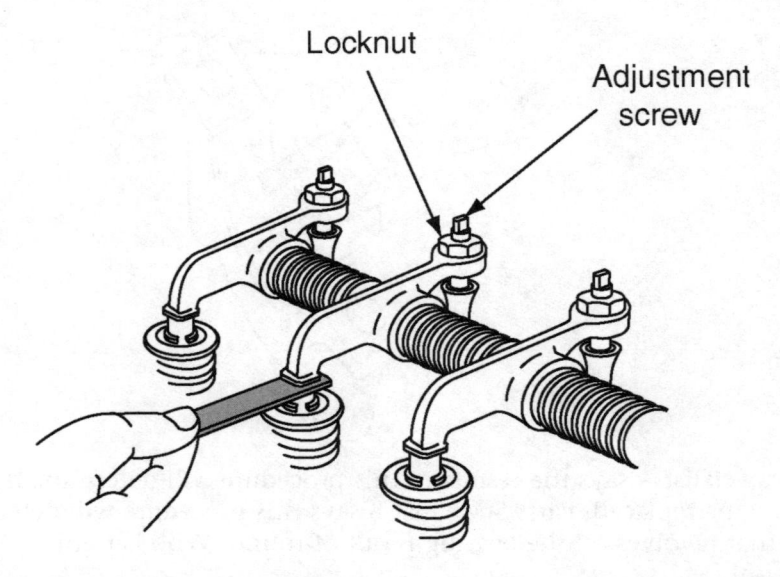

19. Parts Specialist A says the procedure shown will determine if the rocker arms need to be replaced. Parts Specialist B says the procedure adjusts valve lash. Who is right?
    A. A only
    B. B only
    C. Both A and B
    D. Neither A nor B

(C.1.2)

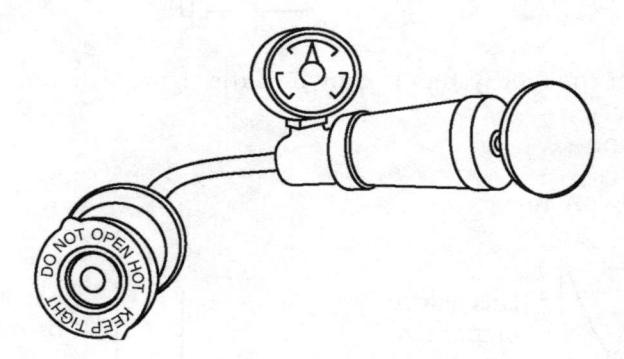

20. Which of the following is not an accurate statement about the use of the tester shown?
    A. The tester is used to create a vacuum on the radiator cap.
    B. The tester creates pressure on the radiator cap.
    C. If the radiator cap is bad, it will not allow pressure to build up.
    D. Every radiator cap is designed to open at a particular pressure.

(C.2.4)

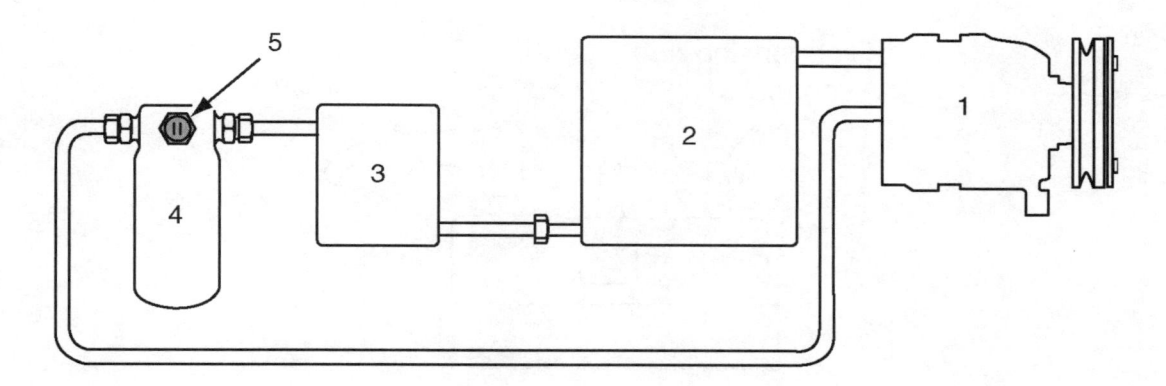

21. Parts Specialist A says the component labeled 3 is the radiator. Parts Specialist B says the component labeled 1 is the air pump. Who is right?
    A. A only
    B. B only
    C. Both A and B
    D. Neither A nor B

    (C.12.1)

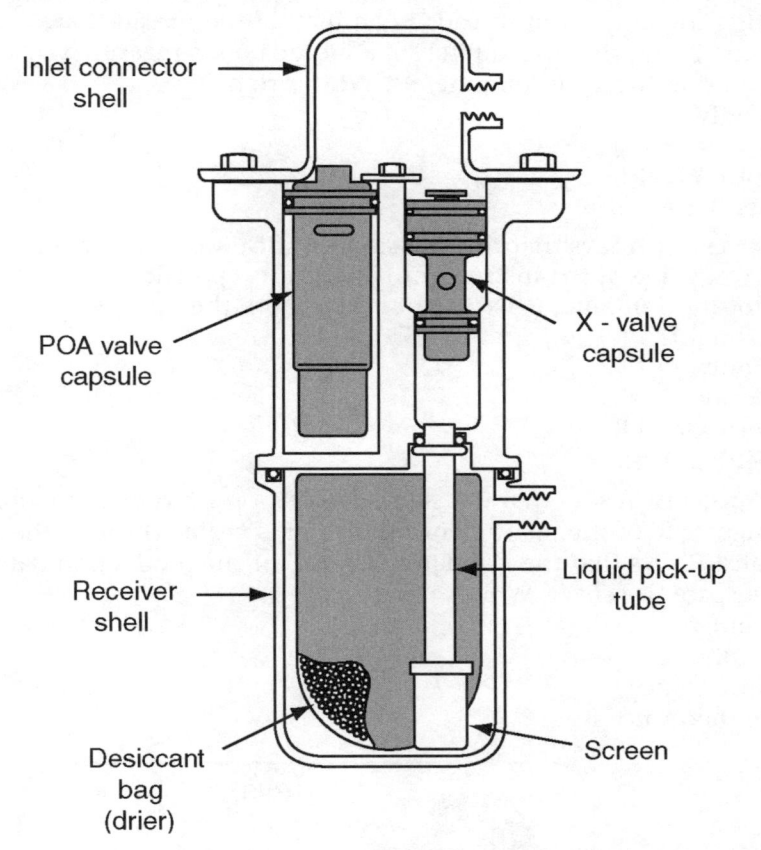

22. While discussing the unit shown, Parts Specialist A says the desiccant bag in this unit is replaceable. Parts Specialist B says this unit controls refrigerant flow through the evaporator and removes moisture from the refrigerant. Who is right?
    A. A only
    B. B only
    C. Both A and B
    D. Neither A nor B

    (C.12.2)

High tension connector

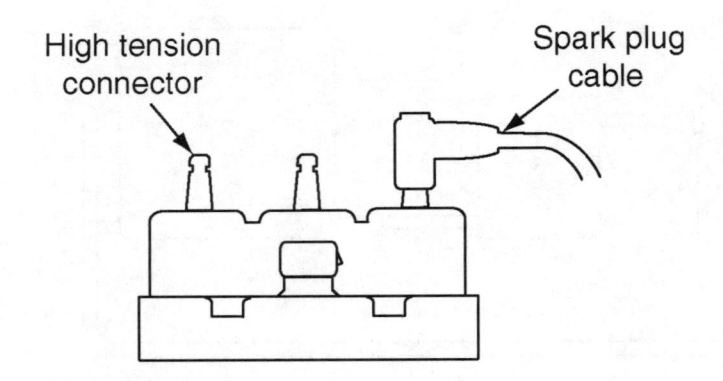

Spark plug cable

23. The component shown is part of what system?
    A. The ignition system
    B. The fuel injection system
    C. The brake system
    D. The air conditioning system (C.4.2)

24. Parts Specialist A says the thread pitch of a fractional bolt is determined by counting the number of threads in an inch of the threaded area of the bolt. Parts Specialist B says the thread pitch of a metric bolt is measured by the distance between the threads in millimeters. Who is right?
    A. A only
    B. B only
    C. Both A and B
    D. Neither A nor B (C.14.3)

25. Part Specialist A says that the vehicle identification number (VIN) is attached to the driver's side of the instrument panel. Parts Specialist B says that the vehicle identification number (VIN) is not visible from the outside of the vehicle. Who is right?
    A. A only
    B. B only
    C. Both A and B
    D. Neither A nor B (D.1)

26. Parts Specialist A says that the standard location for the production date is on the passenger side of the instrument panel that is visible through the windshield. Parts Specialist B says that the standard location for the production date is in the trunk on the spare tire cover. Who is right?
    A. A only
    B. B only
    C. Both A and B
    D. Neither A nor B (D.2)

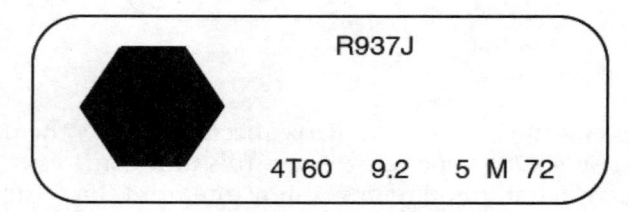

R937J

4T60 9.2 5 M 72

27. Parts Specialist A says that the identification (ID) tag shown in the figure could be from an axle. Parts Specialist B says instead of using a tag, the information could have been stamped into the axle housing. Who is right?
    A. A only
    B. B only
    C. Both A and B
    D. Neither A nor B (D.3)

28. Parts Specialist A says that to find the paint code one can look on the service part identification label. Parts Specialist B says that the service parts identification label can be mounted under the hood or on the driver's door. Who is right?
    A. A only
    B. B only
    C. Both A and B
    D. Neither A nor B   (D.6)

29. Parts Specialist A says that most catalogs have dates on them. Parts Specialist B says that all catalogs are the same size so they will fit a standard binder. Who is right?
    A. A only
    B. B only
    C. Both A and B
    D. Neither A nor B   (E.1)

30. A footnote is used for:
    A. locating special parts.
    B. referencing additional information.
    C. checking inventory.
    D. checking the back order list.   (E.2)

31. Correction bulletins are used to:
    A. update the catalog for new parts releases.
    B. update outdated information.
    C. correct inventory problems.
    D. make corrections during catalog maintenance.   (E.3)

Distance Sensor	Vehicle Speed Sensor	VSS
Distributor Ignition	Distributor Ignition	DI
Distributorless Ignition	Electronic Ignition	EI
DLC (Data Link Connector)	Data Link Connector	DLC
DLI (Distributorless Ignition)	Electronic Ignition	EI
DS (Detonation Sensor)	Knock Sensor	KS
DTC (Diagnostic Trouble Code)	Diagnostic Trouble Code	DTC
DTM (Diagnostic Test Mode)	Diagnostic Test Mode	DTM
Dual Bed	Three Way + Oxidation Catalytic Converter	TWC + OC
Duty Solenoid for Purge Valve	Evaporative Emission Canister Purge Valve	EVAP Canister Purge Valve
E2PROM (Electrically Erasable Programmable Read Only Memory)	Electrically Erasable Programmable Read Only Memory	EEPROM
Early Fuel Evaporation	Early Fuel Evaporation	EFE
EATX (Electronic Automatic Transmission/ Transaxle)	Automatic Transmission	A/T
	Automatic Tranaxle	A/T
EC (Engine Control)	Engine Control	EC
ECA (Electronic Control Assembly)	Powertrain Control Module	PCM
ECL (Engine Coolant Level)	Engine Coolant Level	ECL
ECM (Engine Control Module)	Engine Control Module	ECM
ECT (Engine Coolant Temperature)	Engine Coolant Temperature	ECT
ECT (Engine Coolant Temperature) Sender	Engine Coolant Temperature Sensor	ECT Sensor
ECT (Engine Coolant Temperature) Sensor	Engine Coolant Temperature Sensor	ECT Sensor
ECT (Engine Coolant Temperature) Switch	Engine Coolant Temperature Switch	ECT Switch

32. In the figure, the abbreviation DLC means:
    A. Data Link Connector.
    B. Digital Link Connector.
    C. Diagnostic Link Connector.
    D. Distributorless Link Connector.   (E.4)

33. Parts Specialist A says that car batteries are more likely to fail in the winter season so more batteries should be ordered and in stock for the winter season. Parts Specialist B says in the summer months it is not necessary to stock windshield washer fluid because no sales will be made. Who is right?
    A. A only
    B. B only
    C. Both A and B
    D. Neither A nor B   (G.4)

34. To keep shelf appearance and condition good, shelves should be:
    A. rebuilt once a year.
    B. painted bright colors.
    C. labeled.
    D. taller than six feet.   (G.3)

35. Parts Specialist A says that a parts specialist sometimes will post the cost of the item on a brightly colored poster board to attract the attention of customers. Parts Specialist B says that the display by itself should be enough to let customers know what is on sale. Who is right?
    A. A only
    B. B only
    C. Both A and B
    D. Neither A nor B   (G.2)

36. Parts Specialist A says that the parts specialist should build a display with the product being sold because this will be appealing to customers and it will act as an incentive to buy. Parts Specialist B says that the parts specialist should build the display as tall as possible so that it can be seen wherever you are in the store or department. Who is right?
    A. A only
    B. B only
    C. Both A and B
    D. Neither A nor B   (G.1)

37. A customer orders six liters of transmission fluid. How many quarts should the parts specialist give to the customer?
    A. 2
    B. 4
    C. 5
    D. 6   (A.3)

PORT 2	PARTS-J3	INVOICING		10OCT00 1337

```
              Invoice : 12223                              Sale Type: CASH
             Customer: 99999
                 Name: CASH RETAIL                         Parts:          63.20
              Address:                                     Freight:         0.00
          City,St Zip:                                     Tax:             3.79
           Home Phone:                                     Total Invoice:  66.99
                                                           Backorder Amount: 0.00

 Emp: 117        Sales Person:-         Ship Via: carry        PO: none        B/L:-
```

Part No.	Description	Bin	O.H.	Cost	Sale	Ext. sale	Q.S.	# A	O.O.	PM
23196550	TIMING BELT	245	11	26.73	37.42	37.42	1			
74073158	BELT, SERP	213	3	11.46	16.04	16.04	1			
21646146	PAPER RAGS	218	8	3.52	4.93	4.93	1			
7148887	DRESSING	13B	5	2.19	3.07	3.07	1			
R23895	TAPE	29H	1	1.24	1.74	1.74	1			M

```
      F1=Help              F3=Save      F4=Cancel          F8=Print         F10=login
```

38. According to the invoice, how much did the customer pay for rags?
    A. $2.19
    B. $3.07
    C. $3.52
    D. $4.93   (A.6)

39. Parts Specialist A says that charge accounts are for people who do not have the money at the time of their purchase. Parts Specialist B says that a parts specialist must be familiar with the store's charge policies before making a charge transaction. Who is right?
    A. A only
    B. B only
    C. Both A and B
    D. Neither A nor B   (A.7)

40. Parts Specialist A says that once a parts specialist has been trained, he or she should be on their own and be capable of performing their job responsibilities without assistance. Parts Specialist B says an experienced parts specialist may be required to assist in training a new employee to become a parts specialist. Who is right?
    A. A only
    B. B only
    C. Both A and B
    D. Neither A nor B   (A.10)

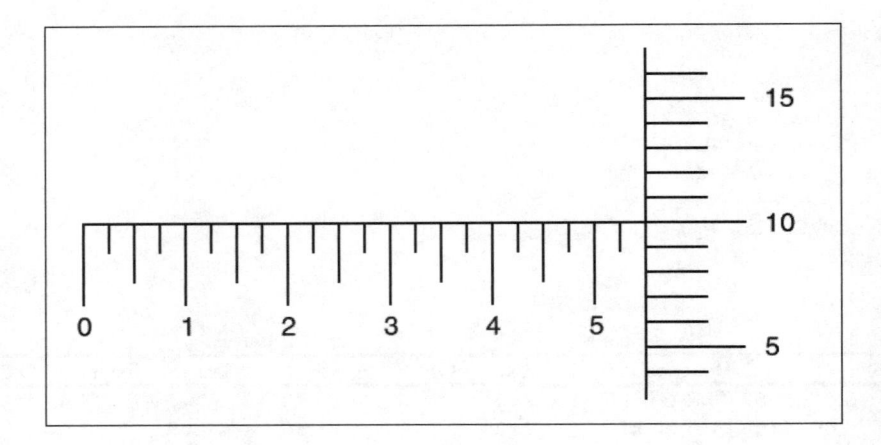

41. Parts Specialist A says the micrometer reading in the figure is 0.560 mm. Parts Specialist B says the micrometer reading in the figure is 0.560 inch. Who is right?
    A. A only
    B. B only
    C. Both A and B
    D. Neither A nor B (A.5)

42. A parts specialist is responsible for keeping all of the following items clean and orderly **EXCEPT:**
    A. floors.
    B. shelves.
    C. shop parts cleaner.
    D. displays. (A.9)

43. Parts Specialist A says that small parts usually do not weigh much, therefore there is no need to seek help when handling them. Parts Specialist B says that knowing the proper way to lift and work within one's ability is important for handling parts. Who is right?
    A. A only
    B. B only
    C. Both A and B
    D. Neither A nor B (A.11)

44. A company that uses hazardous chemicals must follow the guidelines of the:
    A. Environmental Protection Agency (EPA).
    B. Occupational Safety and Health Administration (OSHA).
    C. Department of Public Information Office (DPIO).
    D. Material Safety Data Sheets (MSDS). (A.12)

45. What should be the LEAST-Likely item found near a store exit?
    A. Barrels of oil
    B. 10-pound bags of floor sweep
    C. 6-gallon gasoline cans
    D. Spark plug display (A.13)

46. Parts Specialist A says that a parts specialist should always sell the customers what they want even if it will not fix the problem. Parts Specialist B says that only parts that are needed for a particular repair should be sold to the customers. Who is right?
    A. A only
    B. B only
    C. Both A and B
    D. Neither A nor B (B.1)

47. Parts Specialist A says that a telephone customer cannot see the parts specialist's body language so all communication must be accomplished with words, tones, timing, and inflection. Parts Specialist B says that a parts specialist must remember to speak clearly and slowly enough to be understood and must request that the customer do so if he or she cannot be understood. Who is right?
    A. A only
    B. B only
    C. Both A and B
    D. Neither A nor B (B.6)

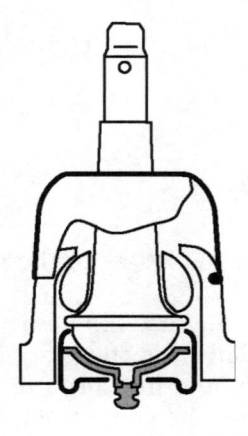

48. While discussing the suspension part shown in the drawing, Parts Specialist A says it is part of the steering linkage and is called a tie-rod end. Parts Specialist B says the part allows the wheel spindle to rotate when the steering wheel is turned. Who is right?
    A. A only
    B. B only
    C. Both A and B
    D. Neither A nor B (C.11.2)

49. Parts Specialist A says that it is important to keep a rag handy for wiping dirt and grime from the counter. Parts Specialist B says that the parts specialist is in charge of the store layout. Who is right?
    A. A only
    B. B only
    C. Both A and B
    D. Neither A nor B (B.8)

Wooden dowel

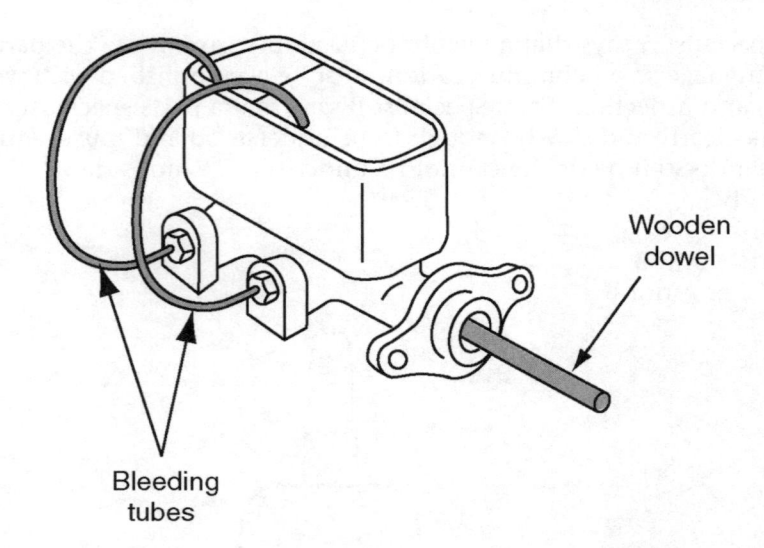

Bleeding tubes

50. Which of the following statements about the procedure shown in the figure is NOT true?
    A. This should be done before installing a new brake master cylinder.
    B. This is done to remove air from the internal pistons in the cylinder.
    C. The cylinder must have fluid in it prior to performing this procedure.
    D. The dowel and tubes must be kept in place during installation. (C.10.4)

51. Parts Specialist A says that front-wheel-drive vehicles have a driveshaft for each front wheel. Parts Specialist B says that the outboard constant velocity (CV) joint is a plunger type. Who is right?
    A. A only
    B. B only
    C. Both A and B
    D. Neither A nor B (C.9.2)

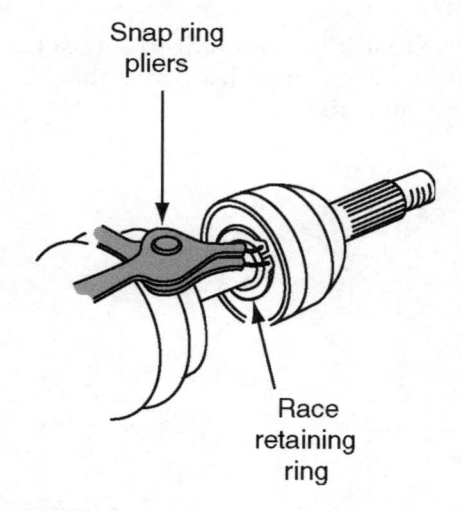

Snap ring pliers

Race retaining ring

52. Parts Specialist A says the part being removed in the drawing is an outboard CV joint. Parts Specialist B says the retaining ring on the part shown should not be reused. Who is right?
    A. A only
    B. B only
    C. Both A and B
    D. Neither A nor B (C.9.4)

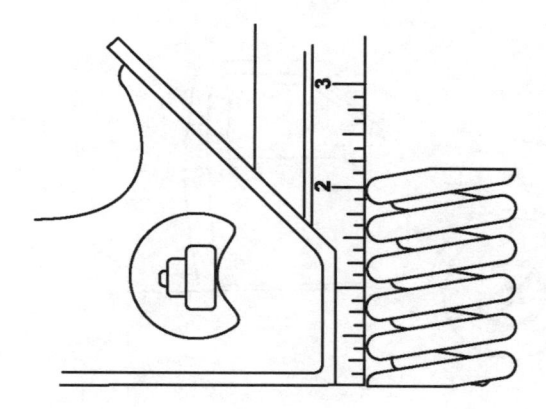

53. Parts Specialist A says if the spring in the figure is distorted, it should be compressed and heated until it is square. Parts Specialist B says if the freestanding height of the spring is lower than specs, the problem should be corrected by the use of shims installed at the spring seat. Who is right?
    A. A only
    B. B only
    C. Both A and B
    D. Neither A nor B (C.11.1)

54. Parts Specialist A says that the vehicle's electrical system is protected by fuses, circuit breakers, and fusible links. Parts Specialist B says that a shorted circuit may blow a fuse or open a circuit breaker or fusible link. Who is right?
    A. A only
    B. B only
    C. Both A and B
    D. Neither A nor B (C.13.2)

55. A bolt with three radial lines embossed on the head would be:
    A. grade 1.
    B. grade 3.
    C. grade 5.
    D. metric. (C.14.4)

56. Parts Specialist A says that the identification (ID) data could be stamped into the casting. Parts Specialist B says that the identification (ID) data is on all of the major components of a vehicle. Who is right?
    A. A only
    B. B only
    C. Both A and B
    D. Neither A nor B (D.3)

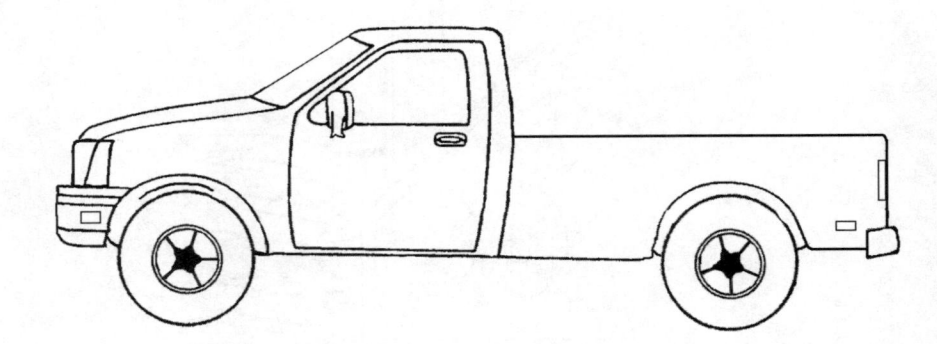

57. Parts Specialist A says that the vehicle in the illustration is a crew cab vehicle. Parts Specialist B says that an extended cab vehicle could have two, three, or four doors. Who is right?
    A. A only
    B. B only
    C. Both A and B
    D. Neither A nor B (D.4)

58. Most catalogs contain all of the following information **EXCEPT**:
    A. form number(s).
    B. date of issue.
    C. the name of product lines and manufacturers.
    D. correction form(s). (E.1)

59. Parts Specialist A says that supersession bulletins may include part numbers that supersede previously issued part numbers. Parts Specialist B says that technical bulletins may alert counter personnel to any unusual installation or application problems. Who is right?
    A. A only
    B. B only
    C. Both A and B
    D. Neither A nor B (E.3)

60. Parts Specialist A says that to verify an abbreviation one should refer to the abbreviation list. Parts Specialist B says that abbreviations are only used in footnotes. Who is right?
    A. A only
    B. B only
    C. Both A and B
    D. Neither A nor B (G.4)

61. Which of the following items is LEAST-Likely to be sold in the summer?
    A. Air conditioning refrigerant
    B. A car battery
    C. Dry gas
    D. Engine coolant (G.4)

62. All of the following statements are true about shelf appearance and condition **EXCEPT**:
    A. Items on a shelf should be clean and new looking.
    B. Returned items cannot be put back on a shelf because they will look used.
    C. Items should be pulled forward to give a full and easy-to-find appearance.
    D. Labels can be used to help organization. (G.3)

63. Which of the following display pricing methods is the LEAST effective method used to promote a product?
    A. Tagging items on the shelf with a price
    B. Setting up a display of the product with an explanation and the price
    C. Displaying prices in the window for potential customers to see
    D. Distributing flyers to encourage product sales (G.2)

64. The LEAST important part of a display is the:
    A. location of the display.
    B. arrangement of the display.
    C. product on the display.
    D. cost of the item on the display. (G.1)

65. An impulse item is:
    A. something a customer needs to fix his or her vehicle.
    B. something a vendor will stock if it can be purchased cheaply.
    C. a flashy item that may catch a customer's eye while browsing and that may lead to a potential sale.
    D. an item needed to make an emergency repair. (G.4)

66. Parts Specialist A says that 327 cubic inches is equal to about 5.8 liters. Parts Specialist B says that 488 cubic inches is equal to 8 liters. Who is right?
    A. A only
    B. B only
    C. Both A and B
    D. Neither A nor B

```
PORT 2              PARTS-J3                INVOICING                                    10OCT00 1337

        Invoice : 12223                                         Sale Type: CASH
       Customer: 99999
           Name: CASH RETAIL                                         Parts:       165.94
        Address:                                                   Freight:         0.00
      City,St Zip:                                                     Tax:         9.96
     Home Phone:                                               Total Invoice:     175.90
                                                            Backorder Amount:       0.00

 Emp: 117        Sales Person:-          Ship Via: carry            PO: none            B/L:-
 Part No.       Description    Bin   O.H.   Cost   Sale   Ext. sale  Q.S.    # A    0.0.    PM
     23196550  CAMSHAFT        245    11    89.18  124.85   124.85 1
     74073158  NUT             213     3     0.34    0.48     0.95 2                  M
     21646146  PAPER RAGS      218     8     3.52    4.93     4.93 1
      7148887  TIMING BELT     13B     5    22.54   31.56    31.56 1
      R23895   OIL SEAL        29H    22     2.61    3.65     3.65 1

     F1=Help         F3=Save        F4=Cancel       F8=Print           F10=login
```

67. While discussing the information on the invoice, Parts Specialist A says the markup on the nut is sixty-one cents. Parts Specialist B says the markup on the camshaft is 40 percent. Who is right?
    A. A only
    B. B only
    C. Both A and B
    D. Neither A nor B (A.6)

68. Parts Specialist A says that a purchase order is used to purchase parts for the store. Parts Specialist B says to only make an emergency order if the vehicle does not have to be worked on right away. Who is right?
    A. A only
    B. B only
    C. Both A and B
    D. Neither A nor B (A.14)

69. There are four different variations of brake shoes in stock that fit the customer's vehicle. Parts Specialist A says that no matter how many different kinds of brake shoes the store carries, it is still the customer's decision. Parts Specialist B says that a parts specialist should always explain the benefits and cost of each kind of brake shoe. Who is right?
    A. A only
    B. B only
    C. Both A and B
    D. Neither A nor B (B.10)

70. A parts specialist is serving a customer at the counter and the phone rings. The parts specialist must put the telephone customer on hold. How should the parts specialist handle this situation?
    A. Answer the telephone, identify the business, and politely say, "Please hold one moment."
    B. Just push the hold button.
    C. Answer the telephone and serve the phone customer while the counter customer waits.
    D. Yell for someone to pick up the telephone. (B.12)

71. Parts Specialist A says that making related sales helps build profits. Parts Specialist B says that making related sales helps the customer as well as the store. Who is right?
    A. A only
    B. B only
    C. Both A and B
    D. Neither A nor B (B.14)

72. Parts Specialist A says that product additions can be found in the table of contents of any parts catalog. Parts Specialist B says that, in most cases, the index will not help in finding a part in a parts catalog. Who is right?
    A. A only
    B. B only
    C. Both A and B
    D. Neither A nor B (E.5)

73. Parts Specialist A says that an index will not indicate if the catalog is for passenger cars or trucks. Parts Specialist B says that parts illustrations can be found by using the table of contents. Who is right?
    A. A only
    B. B only
    C. Both A and B
    D. Neither A nor B (E.5)

74. What information does a packing list contain?
    A. The type of vehicle the delivery person is driving
    B. Where the delivery person will be delivering the shipment
    C. The selling store's business hours
    D. A parts list and how many items are in the shipment (F.2)

75. Parts Specialist A says that a packing list should be included with each shipment. Parts Specialist B says that one packing list can contain more than one part. Who is right?
    A. A only
    B. B only
    C. Both A and B
    D. Neither A nor B   (F.2)

76. Parts Specialist A says that a physical inventory should be done once a year. Parts Specialist B says that keeping good inventory will help ensure that parts are in stock for the customer when needed. Who is right?
    A. A only
    B. B only
    C. Both A and B
    D. Neither A nor B   (F.3)

77. After the physical inventory has been completed, the parts specialist should:
    A. fill out a back order form.
    B. reload the computer software.
    C. report all discrepancies.
    D. rotate the stock.   (F.4)

78. Parts Specialist A says that all of the discrepancy forms should be kept and filed. Parts Specialist B says that after a discrepancy form has been filled out, it must be corrected in the computer system. Who is right?
    A. A only
    B. B only
    C. Both A and B
    D. Neither A nor B   (F.4)

79. Stock should be rotated:
    A. once a year.
    B. every time the shelves are stocked.
    C. every month.
    D. weekly.   (F.5)

80. All of the following parts may have a core charge **EXCEPT:**
    A. brake shoes.
    B. brake calipers.
    C. water pumps.
    D. spark plugs.   (F.7)

81. A warranty-return part should be:
    A. stored in the same room as good parts.
    B. sent to other stores to show the poor quality.
    C. be kept in the delivery truck for disposal.
    D. kept off the shelf, not to be sold again.   (F.8)

82. A pair refers to:
    A. 1.
    B. 2.
    C. 4.
    D. 6.   (F.9)

83. Each refers to:
    A. 10.
    B. 5.
    C. 2.
    D. 1.   (F.9)

84. Parts Specialist A says that a restocking charge could be placed on an item that is returned for a refund. Parts Specialist B says that freight charges are only added to stock orders. Who is right?
    A. A only
    B. B only
    C. Both A and B
    D. Neither A nor B (F.10)

85. The main benefit of using a lost sales report is:
    A. to keep track of the parts that were stolen by shoplifters.
    B. to keep track of the parts that were sold to professional shops.
    C. to keep track of parts that are selling and those that are not selling.
    D. to keep track of parts that were sold to do-it-yourselfers. (F.1)

LEE'S AUTOMOTIVE
www.leesezparts.com

Date: ___ / ____ / _____

Customer: _____

Customer Type: _____ Account #: _____

Counterperson: _____

Part Manufacturer: _____

Part Number: _____

Reason for Lost Sale: _____

_____

86. What type of report is shown in the figure?
    A. A stock order
    B. A shoplifting report
    C. A special order
    D. A lost sales report (F.1)

87. The inside of the air cleaner is contaminated with engine oil. Parts Specialist A says the positive crankcase ventilation (PCV) valve clean air filter in the air cleaner may be plugged. Parts Specialist B says the hose from the positive crankcase ventilation (PCV) valve to the intake manifold may be severely restricted. Who is right?
    A. A only
    B. B only
    C. Both A and B
    D. Neither A nor B (C.6.3)

88. Repeated extension housing seal failure may be caused by:
    A. a scored driveshaft yoke.
    B. excessive output shaft end play.
    C. excessive input shaft end play.
    D. a worn output shaft bearing. (C.7.3)

89. The fluid in an automatic transaxle is a dark brown color and has a burned smell. Parts Specialist A says this problem may be caused by a worn front planetary sungear. Parts Specialist B says this problem may be caused by worn friction-type clutch plates. Who is right?
    A. A only
    B. B only
    C. Both A and B
    D. Neither A nor B   (C.8.3)

90. Parts Specialist A says a worn U-joint may cause a squeaking noise that decreases in relation to vehicle speed. Parts Specialist B says a heavy vibration that only occurs during acceleration may be caused by a worn centering ball and socket on a double Cardan U-joint. Who is right?
    A. A only
    B. B only
    C. Both A and B
    D. Neither A nor B   (C.9.3)

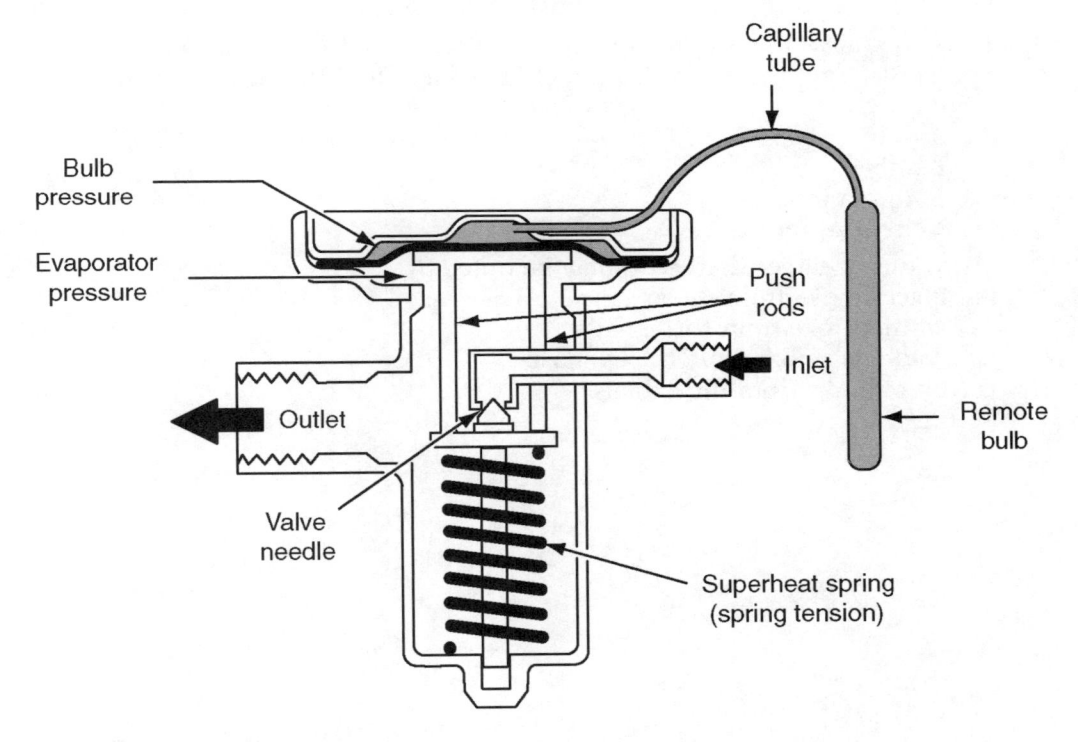

91. The air conditioning part shown in the figure is a(n):
    A. accumulator.
    B. thermal expansion valve.
    C. receiver/drier.
    D. low-pressure relief valve   (C.12.2)

92. Parts Specialist A says that an ohmmeter may be connected to a circuit in which current is flowing. Parts Specialist B says when testing a spark plug wire with a 20,000 Ohm resistance, use the ×100 meter scale. Who is right?
    A. A only
    B. B only
    C. Both A and B
    D. Neither A nor B   (C.13.3)

93. Parts Specialist A says that a gasoline engine is an internal combustion engine. Parts Specialist B says that a diesel engine is an internal combustion engine. Who is right?
    A. A only
    B. B only
    C. Both A and B
    D. Neither A nor B (C.1.1)

94. Flywheel run out is measured with the use of a:
    A. torque gauge.
    B. feeler gauge.
    C. micrometer.
    D. dial indicator. (C.1.3)

95. Blue exhaust is an indication of:
    A. a sound engine.
    B. excessive oil consumption.
    C. rich air/fuel mixture.
    D. coolant leaking into the combustion chamber. (C.1.4)

96. Parts Specialist A says that a missing air filter can affect engine performance. Parts Specialist B says that a restricted fuel filter can affect engine performance. Who is right?
    A. A only
    B. B only
    C. Both A and B
    D. Neither A nor B (C.3.2)

97. Worn turbocharger shaft seals may be noted by:
    A. black smoke from the exhaust.
    B. white smoke from the exhaust.
    C. colorless smoke from the exhaust.
    D. blue smoke from the exhaust. (C.3.3)

Ohmmeter

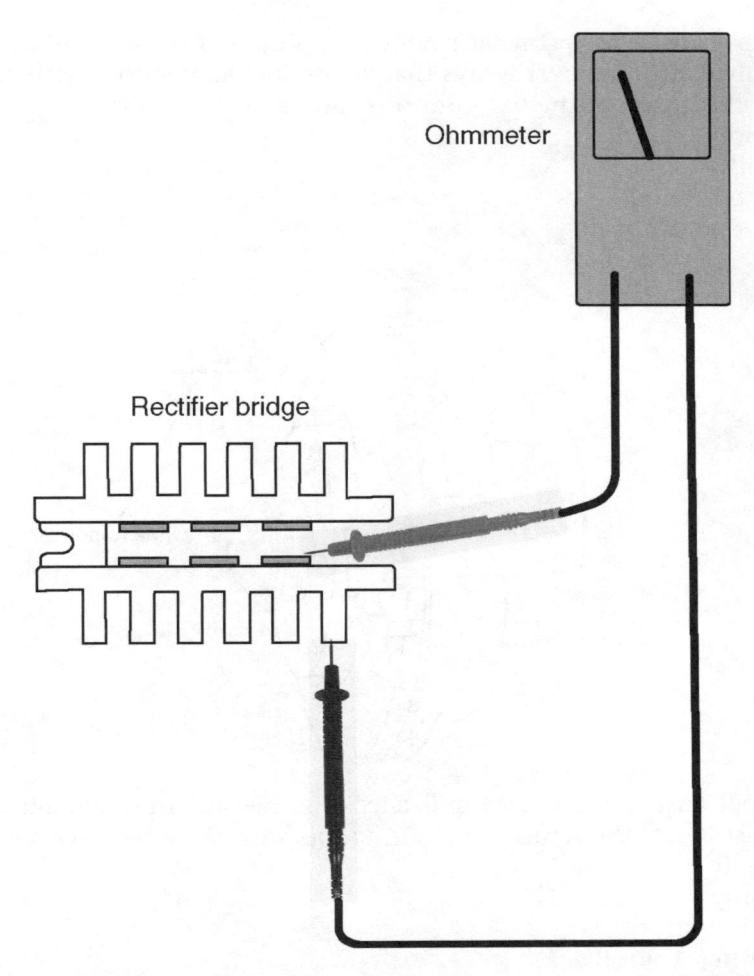

Rectifier bridge

98. Which of the following statements about the test shown in the illustration is not true?
    A. The meter hookup is checking a diode.
    B. The meter should read high resistance if the component is good.
    C. The meter should show zero resistance.
    D. The meter should read low resistance if the component is good. (C.13.4)

99. Which of the following is not found on all vehicles?
    A. Muffler
    B. Exhaust manifold
    C. Exhaust pipe
    D. Resonator (C.5.2)

100. The evaporative emission control system:
    A. introduces exhaust gases into the intake air to reduce the formation of oxides.
    B. breaks down oxides of nitrogen into nitrogen and oxygen.
    C. introduces fresh air into the exhaust stream to cause a second burning.
    D. draws vapors from the fuel tank and introduces them into the intake air stream. (C.6.1)

101. Driveshafts may be made of all the following materials **EXCEPT:**
    A. steel tubing.
    B. aluminum tubing.
    C. composite
    D. titanium. (C.9.2)

102. Parts Specialist A says constant velocity (CV) joints are serviced as a complete assembly. Parts Specialist B says that grease included with a repair kit need not be used if the grease in the old joint was sufficient. Who is right?
    A. A only
    B. B only
    C. Both A and B
    D. Neither A nor B

(C.9.3)

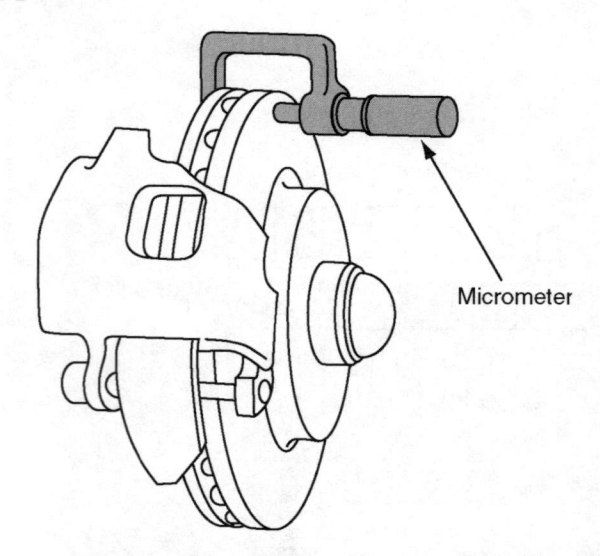

103. Parts Specialist A says the setup is checking the run out of a brake rotor. Parts Specialist B says the setup is measuring the rotor thickness. Who is right?
    A. A only
    B. B only
    C. Both A and B
    D. Neither A nor B

(C.10.4)

104. Parts Specialist A says that bolt diameter is the measure across the threaded area. Parts Specialist B says that a thread pitch gauge is used to determine thread pitch. Who is right?
    A. A only
    B. B only
    C. Both A and B
    D. Neither A nor B

(C.14.3)

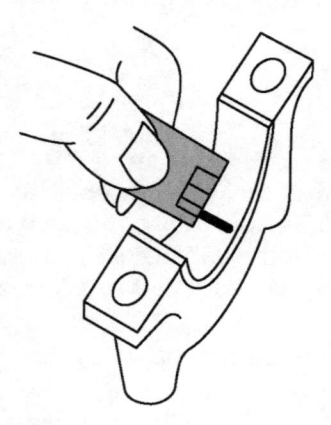

105. The tool shown in the figure is called:
    A. a radius gauge.
    B. Plastigage™.
    C. a bearing scraper.
    D. a concentricity gauge.

(C.1.4)

7 Appendices

Answers to the Test Questions for the Sample Test Section 5

1.	C	27.	A	53.	D	79.	A
2.	B	28.	C	54.	D	80.	B
3.	D	29.	C	55.	B	81.	A
4.	A	30.	D	56.	A	82.	A
5.	B	31.	B	57.	C	83.	B
6.	C	32.	C	58.	C	84.	C
7.	A	33.	C	59.	A	85.	A
8.	C	34.	A	60.	A	86.	A
9.	C	35.	D	61.	D	87.	A
10.	C	36.	D	62.	C	88.	B
11.	C	37.	A	63.	D	89.	D
12.	B	38.	D	64.	B	90.	A
13.	B	39.	C	65.	C	91.	D
14.	D	40.	A	66.	C	92.	B
15.	B	41.	D	67.	B	93.	D
16.	D	42.	B	68.	D	94.	C
17.	D	43.	C	69.	B	95.	B
18.	D	44.	B	70.	C	96.	C
19.	A	45.	D	71.	A	97.	A
20.	C	46.	B	72.	C	98.	B
21.	C	47.	B	73.	A	99.	A
22.	A	48.	A	74.	A	100.	C
23.	C	49.	A	75.	C	101.	B
24.	C	50.	C	76.	D	102.	D
25.	C	51.	C	77.	C		
26.	D	52.	D	78.	A		

Explanations to the Answers for the Sample Test Section 5

Question #1
Answer C is correct in our $10.00 part example. Both parts specialists are correct. A part purchased for $10.00 and sold for $15.00 generates a 33.3 percent profit margin. Profit margin is the percentage of the original part cost you have left over after you pay for its cost. Also, a 10 percent discount on a $25.00 sale is $2.50. This is an expression of mark-up or in this case mark down. Mark-up is simply a percentage of the sale price. This is not the same as profit margin.

Question #2
Answer A is wrong, a 5 percent restocking fee would not cost the customer $95.00. A 5 percent restocking fee would cost the customer $5.00; therefore B is right. C is wrong, $100.00 is what the part cost. D is also wrong; the customer paid $100.00 and would not receive more money back than was paid.
Answer B is correct.

Question #3
Answer D is correct. Both parts specialists are wrong; 350 cubic inches is about 5.7 liters and 390 cubic inches is about 6.4 liters.

Question #4
Answer A is correct. Only choice A is right. 369482A is before all of the other numbers. 378654C is after 369482A, 369482B is after 369482A, and 388426A is after 369482A.

Question #5
Answer B is correct. The micrometer reading is 0.184 inch.

Question #6
Answer C is correct. Both parts specialists are right. You must be familiar with the charge policies before making a charge transaction. Also, charge accounts are typically for large sales.

Question #7
Answer A is correct, since only Parts Specialist A is right. The ability to communicate with others is essential for anyone in any business. Parts Specialist B is wrong; it can be very difficult to interact with some people. This is why effort must be made to be patient and understanding, regardless of the customer's behavior.

Question #8
Answer C is correct. Both specialists are correct. Although the store or parts department may have an employee or cleaning crew that have the responsibility for cleaning the place, a counterperson also has the responsibility for keeping the place tidy and professional looking.

Question #9
Answer C is correct. Both parts specialists are right. An experienced employee is often asked to help train a new employee. It is also better to ask for help than to sell the customer the wrong part.

Question #10
Answer A is wrong. Answer A is the sale price of the motor. This is the amount that will be reimbursed to the customer plus the tax paid on that amount.
Answer B is also wrong; this is the store's recorded cost for the motor.
Answer C is correct. This is the sale price with tax added. The customer should receive this amount.
Answer D is the store's cost plus tax. To totally understand the answer, you may want to calculate the tax rate. To do this, divide the tax paid by the total of the parts (9.50 divided by 158.40). You will find the tax to be 6%. Then if you add 6% to the sale price of the motor, you will have a total of $92.54.

Question #11
Answer C is correct. Always evaluate the task before attempting to lift an object. If the part is too heavy, then ask for assistance. Answers A, B, and D are wrong. You should always evaluate the task before asking for assistance. If the part is too heavy, then ask for help and be willing to help others lift a heavy load. Also, before you go ahead and use a hand truck, make sure the part is too heavy for lifting.

Question #12
Answer B is correct. The EPA and OSHA have strict regulations on solvents, cutting oils, and caustic cleaning compounds. There are no EPA or OSHA regulations on normal floor soaps.

Question #13
Answer B is correct. Parts Specialist A is wrong; large items should not be stored on the top shelf due to the chance that the item could fall and injure someone. Parts Specialist B is right; expensive items should be displayed behind the counter in locked cases to discourage theft.

Question #14
Answer D is correct. Do-it-yourselfers often need good advice on parts and methods. Do-it-yourselfers usually need more information than professional automotive customers, but they do not always require technical assistance. They usually need good advice on parts and methods but do not always purchase used or rebuilt parts. They also usually need good advice on parts and methods, but seldom do they attempt major engine or transmission repairs.

Question #15
Answer B is correct. The tester shown in the figure is a standard cooling system tester. It is used to check for leaks throughout the system, including the radiator cap. The tester is used to create a high pressure in the system or on the radiator cap. If there is a leak, the pressure will not build up. In the case of a radiator cap, pressure should be able to build up to the pressure rating of the cap. If the cap cannot hold a lower pressure, it is bad and should be replaced. The tester does not affect the temperature; therefore, only Parts Specialist B is right.

Question #16
Answer D is correct. Both parts specialists are wrong. A parts specialist's opening question should be designed to put the customer at ease and elicit as much data as possible. Greeting with the question, "May I help you?" gives the customer an easy opportunity to reply, "No."

Question #17
Answer D is correct because both parts specialists are wrong. A parts specialist should not be concerned with placing blame because it is not always evident that there is blame to place. Also, a parts specialist should take whatever time necessary to understand the customer's viewpoint and should never raise his or her voice.

Question #18
Answer D is correct. Both parts specialists are wrong. The first step toward making a customer feel noticed is to make eye contact. Do this when greeting the customer and when talking with him or her. From the time a customer walks through the door until leaving the store, the person should be acknowledged and responded to.

Question #19
Answer A is correct. Parts Specialist A is right; the parts specialist should politely ask the person at the counter to wait while the parts specialist answers the phone. Keep the phone conversation as brief as possible and continuously acknowledge the person at the counter.

Question #20
Answer A is wrong. To measure valve spring free height, the spring is removed from the valve assembly and measured from top to bottom.
Answer B is correct for dimension B, not A. This dimension represents installed valve spring height.
Answer C is correct. Dimension A shows the measurement for installed valve stem height.
Answer D is wrong because there is no true clearance between the retainer and the seat. The spring occupies that distance.

Question #21
Both parts specialists are right and **Answer C is correct.** Customers are generally influenced by a clean, neat, parts specialist and by the condition of the store as well as the appearance of the employees.

Question #22
Answer A is correct. Parts Specialist A is right; selling related parts can often boost profits by as much as 30 percent. In fact, "up-selling" is a characteristic of a great counterperson. Parts Specialist B is wrong; a parts specialist should not force a sale by promoting slow-selling merchandise.

Question #23
Answer C is correct. If you understand the major components of an A/C system and look carefully at the arrangement of the parts, you should be able to identify the parts. Plus as a hint, the drawings give you an outline of the shape of the component. In the drawing, component 1 is the compressor, 2 is the condenser, 3 is the evaporator, 4 is the accumulator, and 5 is a pressure switch. Both parts specialists are correct.

Question #24
Answer C is correct. The component is a VIR. If you look carefully you will note there is a desiccant bag like a typical receiver drier, but there are also control valves in the unit. There is a TXV and a POA valve. This is why it is called a valves-in-receiver unit.

Question #25
Answer C is correct. The part shown is an ignition coil for a distributorless ignition system. Since there are three terminals showing, this coil pack will serve at least three cylinders. Perhaps the terminals for the other cylinders are not noticeable in the drawing as there are few three-cylinder engines. Therefore, both parts specialists are correct.

Question #26
Answer D is correct. This is a tricky question, or so it seems. You are asked to identify the part that is not part of the fuel system. The fuel pump, fuel tank, and carburetor are part of the fuel system. The fuel tank hangers help support and secure the fuel tank in the vehicle but are not a part of the fuel system.

Question #27
Answer A is correct. Parts Specialist A is right; The EGR (exhaust gas recirculation) valve is part of the emissions control system. It recirculates a small amount of exhaust into the combustion chambers to control the temperatures that are reached. Reducing the temperature reduces the amount of NO_x in the exhaust.

Question #28
Answer C is correct. The figure is that of a typical EGR valve. The EGR is a major component of the emission control system in most vehicles.

Question #29
Answer C is correct. The short in the drawing is after the load. It completes the circuit for the lamp causing it to be on always. The switch serves the function of only completing the circuit to ground; therefore, with the short to ground the switch would be useless. A short after the load would not cause amperage to increase and therefore the fuse would not blow, nor would the wire melt or burn. Since there is not an increase in amperage, the switch's contacts would not arc.

Appendices Explanations to the Answers for the Sample Test Section 5

Question #30
Answer D is correct. The meter shown is connected across the ground circuit of a starter motor. This hookup would display the voltage drop across the ground. In an ideal world, this reading should be zero. A voltage drop of 0.2 volts would be slightly high and would indicate excessive resistance in the starter ground circuit. Both specialists are wrong.

Question #31
Answer B is correct. The part shown is a ball joint. A ball joint allows the control arms to move up and down which keeps the tires on the road.

Question #32
Answer C is correct. Both parts specialists are right. A vehicle's suspension system helps to cushion the vehicle from road shocks and helps to keep the vehicle in control of the road or driving surface.

Question #33
Answer C is correct. Both specialists are correct. Camber is a tire wearing angle. Camber changes at the level or height of the suspension changes. Specifications for camber provide good tire contact as the suspension responds to the road surfaces. If the strut is damaged, camber will be wrong.

Question #34
Answer A is correct because Parts Specialist A is right; the starter motor turns a flywheel that is mounted on the rear of the crankshaft. Parts Specialist B is wrong; when the ignition key is turned to the START position, battery voltage is applied directly to the starter motor, not to the alternator.

Question #35
Answers A, B, and C are wrong.
Answer D is correct. Bolt dimensions are determined by the diameter and pitch of the bolt threads. The bolt length is also used in stating dimensions. The distance across the flats of a bolt head determines the size wrench to use; likewise, the correct wrench size to use is determined by the distance across the flats of a bolt head. C is wrong because the distance across the points of a hex head have nothing to do with a bolt's thread diameter.

Question #36
Answer D is correct. Both specialists are wrong. The standard location for the vehicle identification number is attached to the driver side of the instrument panel, visible through the windshield. Many older vehicles had their *serial number* located on inner fender panel.

Question #37
Answer A is correct. The procedure is called bench bleeding a master cylinder. This procedure is recommended prior to installation of the unit to a vehicle. Bench bleeding removes the air from the internal parts of the master cylinder. Failure to do this will make bleeding of the brakes, after installation, very difficult. It is not a substitute for regular brake bleeding. Anytime the brake system is opened, the entire system needs to be bled. Fluid is put in the master cylinder's reservoir and the dowel is used to move the pistons in the cylinder. This movement pumps the fluid through the cylinder and pushes the air out. The tubes and dowel must be removed prior to installation. Parts Specialist B is wrong.

Question #38
Answer D is correct. Both parts specialists are wrong. The paint code is located on a separate trim tag and it is not found in the owner's manual. The paint code is not part of the vehicle identification number (VIN). It is located on a trim tag, called the service parts identification label, which is mounted under the hood or on the driver's door of some vehicles.

Question #39
Answer C is correct. Both parts specialists are right. Each catalog has, in the upper right-hand corner, the form number, the date of issue, and the number and date of the catalog it replaces or supplements. The cover also identifies the product line and manufacturer and often provides an area where the jobber can stamp his or her name for the convenience of dealer customers.

Question #40
Answer A is correct. Parts Specialist A is right; footnotes are often very important for getting the correct part for the customer. Footnotes give additional information to determine exceptions or to find alternative parts. Sometimes footnotes are difficult to understand when a number of them are squeezed on a page. The footnotes themselves are important. Footnotes should be referenced systematically in order to fully define those applications that have multiple possibilities.

Question #41
Answer D is correct. Flare, compression, and pipe type fittings are used in automotive service. Answer A is wrong. A tubing type fitting is not used on automobiles. Answer B is also wrong. A steel type fitting is not used on automobiles. Also, choice C is wrong. Neither steel nor tubing type fittings are used in automobiles.

Question #42
Answer B is correct. Bulletins are lists of updated information that manufacturers send to jobbers between issues of their catalogs. They are the primary tool for conducting catalog maintenance, which involves continually updating and revising the catalog racks and making sure counter personnel are using the most up-to-date information from those catalogs. Bulletins are usually one of the following types: (1) new item availability bulletins which list items now in stock; (2) supersession bulletins, which note part numbers that now supersede previously noted numbers; (3) product information bulletins; which contain specific information about a product, such as a manufacturing defect or a unique method of installation. (4) technical bulletins, which alert counter personnel to any unusual installation or fit problems or unique maintenance tips; and (5) correction bulletins, which refer to catalog errors due to mistyping or inaccurately assigned parts numbers. There are no delivery bulletins.

Question #43
Answer C is correct. In the figure, AC is the abbreviation for air conditioning.

Question #44
Answer B is correct. Parts Specialist A is wrong. Generally, all stock should be rotated so other items are not continually pushed to the rear of the shelf. Parts Specialist B is right. Displaying the price on a shelf helps to maintain the overall appearance of the store. The customer does not have to move the merchandise in order to determine its cost, thus maintaining neater shelf displays.

Question #45
Answer D is correct. This is an except-type question. All of the answer choices are true statements about display pricing except D. Displaying an item does not mean that it will always be on sale. Display pricing is often used for introduction of new items and special sale items. The customer is more likely to notice the price and product in a display than on the shelf.

Question #46
Answer A is wrong because there is no regulation on height.
Answer B is correct. A display arrangement in an open area is most effective.
Answer C is also wrong; display arrangements should be placed in the open away from the usual location for the product.
Answer D is wrong. Displays can be set up by anyone.

Question #47
Answer B is correct. Four liters is almost equal to four quarts. One quart is equal to 0.9464 liters, so one quart is almost equal to one liter.

Question #48
Answer A is correct. An invoice should be completed for each sale in the store. These invoices are used to track inventory and customer purchases for billing purposes. A stock order is used by the store to order more stock from the suppliers. A back order refers to merchandise ordered from a supplier but not shipped, due to the supplier being out of stock. Parts Specialist A is right; an invoice should be completed for each and every sale. Parts Specialist B is wrong; a back order refers to merchandise ordered from a supplier but not shipped due to the supplier being out of stock.

Question #49
Answer A is correct. Defensive selling implies a method of selling that protects the interests of the store in response to do-it-yourselfers who are not always clear and methodical in their diagnosis of a problem. These customers will sometimes purchase the wrong part, install it, find that the problem has not been solved, and then try to return the part as defective. Parts that are sold and returned in this manner are so done at the expense of the store. For this reason, many stores have a policy that does not allow the return of parts that have been installed. Most stores require that parts be returned in their original containers. Therefore Parts Specialist A is right. Parts Specialist B is wrong; do-it-yourselfers may not have enough knowledge and diagnostic skills to replace the correct part.

Question #50
Answer C is correct. When the parts specialist carries many different brands of the same product, the difference between them should be explained. If the store carries, for example, four different kinds of brake pads, the parts specialist should identify the material, size, cost, warranty, and name brand differences. The parts specialist should let the customer decide on which part to buy. The parts specialist can make recommendations, but the final decision should be the responsibility of the customer.

Question #51
Answer C is correct. After some experience, a parts specialist will know when a customer is ready to buy and when a customer is there to browse. A parts specialist should not push the sale of any parts that a customer may not need. Part of a parts specialist's job is to be able to identify the different customer types and determine their needs. Both parts specialists are right.

Question #52
Answer D is correct, because both parts specialists are wrong. If a parts specialist must put a telephone customer on hold, he or she should identify the name of the store and politely state something such as, "Please hold one moment." The parts specialist should never pick up the phone and press the hold button without saying anything, and he or she should never push the button before finishing a sentence or phrase. If the parts specialist knows that it will take a while before he or she can talk with the caller, the parts specialist should ask that the caller call back rather than be put on hold for a long period of time.

Question #53
Answer D is correct. A good parts specialist should always promote sales and store services. If a customer asks for some brake pads, the parts specialist should let the customer know that they have a machine shop and can turn his rotors or drums if needed. The parts specialist can also let the customer know that they have brake fluid and other related brake parts in stock to further promote the store sales.

Question #54
Answer D is correct. Both Parts Specialists A and B are wrong. A good parts specialist should always promote sales and store services. If a customer asks for some brake pads, the parts specialist should let the customer know that they have brake fluid and other related brake parts in stock to further promote the store sales. Selling related items is extremely important, not only to build profits, but also to help the customer achieve the safest possible results.

Question #55
Answer B is correct. A parts specialist should help the customer get the right part each and every time. If a customer comes in for a starter while carrying a pair of jumper cables, the parts specialist might ask, "What was your car's problem that leads you to believe that it needs a starter?" The customer may say, "It will not start without jump-starting the battery." The parts specialist should then explain that the battery could be the problem. This way the parts specialist will not only get the sale for the battery but will not have sold the customer a part that did not fix the vehicle.

Question #56
Answer A is correct. Closing a sale means getting a commitment from the customer. Taking the customer's money follows after the commitment. Looking up the correct part occurs before the customer makes a commitment. Handing the customer their change and receipt occurs after taking their money, which follows closing the sale.

Questions #57
Answer C is correct. Both parts specialists are right. A vehicle build sheet can indicate the engine size and the color of the vehicle.

Question #58
Answer C is correct. Both parts specialists are correct. The index and the table of contents will show the catalog sequence. The index however will provide more detailed information on the particular vehicle manufacturer and model, the sequence they are presented in, and the page number where each will be listed.

Question #59
Answer A is correct. Parts Specialist A is right. Bulletins are the primary tool for conducting catalog maintenance, which involves continually updating and revising the catalog racks and making sure counter personnel are using the most up-to-date information from those catalogs.

Question #60
Answer A is correct. As with any delivery, an order for a shipment must be carefully picked, packaged, and documented. A packing list or slip must be included with each shipment. This list contains a detailed description of the items included in the shipment. A separate packing list can be placed in each part of a multiple-part shipment, but often a single comprehensive list is packaged with or secured to the first part of a multiple-part shipment. Parts Specialist A is right; a packing list contains a detailed description of the items included in the shipment. Parts Specialist B is wrong; a packing list contains a detailed description of the items included in the shipment but no directions for where to deliver the package.

Question #61
Answer D is correct. Physical inventory should be done once a year.

Question #62
Answer C is correct. Physical inventory discrepancies are reported by filling out an inventory discrepancy form.

Question #63
Answer D is correct. Parts Specialist A is wrong; stock rotation prevents older stock from sitting at the back of the shelf and eventually having an aged appearance. Parts Specialist B is also wrong. Stock should be rotated every time that the shelves are stocked. Both specialists are wrong.

Question #64
Answer B is correct. A special order is a part that is ordered because the store does not keep the part in stock.

Appendices
Explanations to the Answers for the Sample Test Section 5

Question #65
Answer C is correct. A core charge is a refundable charge that is added when a customer buys a remanufactured part.

Question #66
Answer C is correct. Both specialists are right. Additional forms must usually be completed for warranty-return parts. This allows the manufacturers to determine what is wrong with the part and to credit the store for the return. A warranty-return part should never be placed back in the inventory to be sold to another customer.

Question #67
Answer B is correct. A case refers to the number of parts packaged together and sold as a unit.

Question #68
Answer D is correct. A restocking fee would be charged to a customer for returning a part for a refund.

Question #69
Answer B is correct. Parts Specialist A is wrong; lost sales reports are beneficial to the store. Parts Specialist B is right; the main benefit of using lost sales reports is the ability to spot trends in the aftermarket parts business.

Question #70
Answer C is correct. Both specialists are right. Flywheel run out causes an uneven clutch plate mating surface resulting in clutch grabbing. The pressure plate must be installed in its original position to maintain proper balance.

Question: #71
Answer A is correct. All of the statements about oil pump clearance measurements are true except choice A. If the clearance between the rotors is normal, there is no need to measure the inner rotor diameter. Measuring the clearance between the rotors, the thickness of the inner and outer rotors, and the clearance between the outer rotor and the housing are valid oil pump measurements.

Question: #72
Answer C is correct. Both specialists are right. The water pump bearing may become contaminated by coolant leaking past the pump seal. A defective water pump bearing may cause a growling noise at idle speed.

Question: #73
Answer A is correct. Parts Specialist A is right; an intake manifold vacuum leak may cause a cylinder to misfire at idle speed. Parts Specialist B is wrong; an intake manifold vacuum leak will not cause a cylinder to misfire during acceleration when the manifold vacuum is reduced.

Question: #74
Answer A is correct. Part number 8 is a cable assembly that moves the self-adjuster lever on the star adjuster when the brake shoes expand as the brakes are applied while backing up. Parts Specialist B is wrong. Part number 17 is used to mechanically apply the brakes when the parking brake is applied.

Question: #75
Answer C is correct. In this except-type question, choice C is wrong. The mechanical advance rotates the reluctor ahead of the distributor shaft in the same direction as shaft rotation. The vacuum advance does rotate the pickup plate in the opposite direction to shaft rotation. The vacuum advance does control spark advance in relation to engine load. The mechanical advance does control the spark advance in relation to engine rpm.

Question: #76
Answer D is correct. This is an except-type question. All of the answer choices are true statements about manifold heat control valves except for choice D. A manifold heat control valve stuck in the closed position increases (not reduces) intake manifold temperature. A manifold heat control valve stuck in the closed position causes a loss of engine power. A manifold heat control valve stuck in the open position may cause an acceleration stumble. A manifold, heat control valve improves fuel vaporization in the intake manifold, especially when the engine is cold.

Question: #77
Answer C is correct. This is not a true statement. The part is an outboard CV joint. It is being removed from an axle or half shaft. The joint normally comes with a pack of lubricant that must be inserted over and in the joint during installation. A boot is installed over the joint to keep the lubricant and joint free of moisture and dirt. Contamination will cause premature joint failure.

Question: #78
Answer A is correct. Most linkage adjustments are performed with the gear shift lever in neutral.

Question: #79
Answer A is correct. A five-speed manual transaxle has a growling and rattling noise in third gear only. The cause of the problem must be something that affects only third gear and only affects that gear when the transaxle is operating that gear. Worn or chipped teeth on the third speed gear could result in a growling noise while driving in third gear.
Answers B and C are wrong. Worn dog teeth on the third speed gear may cause hard shifting or jumping out of third gear, but this defect would not cause a growling noise while driving in third gear.
Answer D is also wrong. Worn threads in the third speed blocking ring may cause hard shifting, but this wear would not cause a growling noise in third gear.

Question: #80
Answer B is correct. Parts Specialist A is wrong; since the rear planetary gearset is not turning with the engine running and the vehicle stopped, a defective rear planetary gearset would not cause this noise. Parts Specialist B is right; since the oil pump is turning continually with the engine running, this component may be the cause of the whining noise.

Question: #81
Answer A is correct. Parts Specialist A is right; a misadjusted manual valve shift linkage may cause low fluid pressure that may cause clutch slipping and premature failure. Parts Specialist B is wrong; a misadjusted manual valve shift linkage may cause low fluid pressure that may cause clutch slipping and premature failure.

Question: #82
Answer A is correct. Parts Specialist A is right; a clunking noise while decelerating may be caused by a worn inner drive axle joint. Parts Specialist B is wrong; a worn front wheel bearing usually results in a growling noise while cornering or driving straight ahead.

Question: #83
Answer B is correct. Parts Specialist A is wrong; damaged brake lines should not repaired using compression fittings. Only double flared or ISO connections with suitable unions should be considered. Parts Specialist B is right; a tube-bending tool should be used to make the necessary brake tubing bends without kinking the brake tubing.

Question: #84
Answer C is correct. This is the normal setup for checking spring squareness and freestanding height. If the spring is distorted, it will not sit squarely on the tool.

Question: #85
Answer A is correct. When one side of the bumper is pushed downward with considerable force and then released, the bumper should only complete one free upward bounce if the shock absorber or strut is satisfactory. The other answer choices are wrong.

Question: #86
Answer A is correct. The meter shown is connected across the ground circuit of a solenoid. This hookup would display the voltage drop across the ground. To measure voltage at the starter, the meter should be connected so the positive lead is on the starter battery terminal and the negative lead on the negative post of the battery. To measure the voltage drop across the solenoid, connect the positive lead to the battery terminal at the solenoid and the negative lead at the ground of the starter. Battery voltage is measured by connecting the leads of the meter across the battery.

Question: #87
Answer A is correct, because Parts Specialist A is right. Restricted evaporator refrigerant passages may cause frosting of the evaporator outlet pipe. Parts Specialist B is wrong; this condition may cause lower, not higher, than specified low-side pressures.

Question: #88
Answer A is wrong; two high meter readings indicate a shorted diode.
Answer B is correct. When testing a diode, connect the ohmmeter leads across the diode and then reverse the leads. A good diode will have one high and one low ohmmeter reading. Answer C is wrong. Two low meter readings indicate an open diode.
Answer D is wrong; a meter reading of 0 Ohms and 10 Ohms indicates a defective diode.

Question: #89
Answer D is correct. Both parts specialists are wrong. The best location to attract impulse buying is between the customer's chest and eye level.

Question: #90
Answer A is correct because only Parts Specialist A is right. There is more than one coil in a distributorless ignition system (DIS). Parts Specialist B is wrong; the ignition module synchronizes the coil in relation to the crankshaft position and firing order. The computer, ignition module, and position sensor combine to control spark timing and advance. The computer collects and processes information to determine the ideal amount of spark advance for the particular operating conditions. The ignition module uses crank/cam sensor data to control the timing of the primary circuit in the coils. Remember that there is more than one coil in a distributorless ignition system. The ignition module synchronizes the coil's firing sequence in relation to the crankshaft position and firing order of the engine. Therefore, the ignition module takes the place of the distributor.

Question: #91
Answer D is correct. The question asks you to identify, from the list, the part that is not part of an exhaust system. A muffler, resonator, and catalytic converter are part of a typical exhaust system. A reverberator is not a part of the exhaust system.

Question: #92
Answer B is correct. Parts Specialist B is right; the clutch pressure plate is operated by a clutch release bearing. Parts Specialist A is wrong; pressing the clutch pedal disengages the clutch.

Question: #93
Answer D is correct. The tool shown is a special tool used to separate a refrigerant line on some systems. The tool depresses the positive lock and the line fitting can be separated. Neither specialist is right.

Question: #94
Answer C is correct. Parts Specialist A is right; the refrigerant leaving a compressor is a high-pressure vapor, not a liquid, it becomes a liquid as it passes through the condenser. Parts Specialist B is right. The refrigerant entering the compressor is a low-pressure vapor.

Question: #95
Answer B is correct because an alternator or generator is on all modern vehicles and certainly would not be considered an accessory. A radio and sound system, power seats and windows, and rear window defogger are not necessary and can be considered accessories, although they may be standard equipment on some vehicles.

Question: #96
Answer C is correct. Both parts specialists are right. A nut should be the same grade as the bolt it is used on and all bolts securing a part should be of the same grade.

Question: #97
Answer A is correct. Only a compression fitting uses a ferrule.

Question: #98
Answer B is correct. Parts Specialist A is wrong; not all hoses used in automotive service withstand high pressure. Air intake, vacuum hoses and radiator hoses operate at low or negative pressure. Parts Specialist B is right; some hoses used in automotive service are made of a reinforced synthetic rubber.

Question: #99
Answer A is correct. Parts Specialist A is correct. The open end wrench will grip a bolt on two sides only, while a box end wrench is designed to grip both sides of each corner of a bolt. So in the case of a hex head bolt, the box end wrench will have six points of contact on the head of the bolt. The box end wrench must be the correct size for the bolt head to be effective.

Question: #100
Answer C is correct. The electrical/charging system performs two basic duties, maintaining the battery's state of charge and providing electrical power for the ignition system, air conditioner, heater, lighting, and all electrical accessories. Electrical energy is stored in the battery. This energy is used to supply the power needed to start the motor and other devices on the vehicle. The charging system is responsible for restoring power back to the battery through electrical output from the alternator.

Question: #101
Answer B is correct. The cold cranking amp is the number of amps the battery can deliver at zero degrees to start the engine, not the number of volts. CCA is the rating assigned to batteries to indicate the amperes that a fully charged battery will maintain for 30 seconds at a specified voltage (usually 10.5 volts). Some manufacturers have different specs. Reserve amperage is the rated output of a battery when there is no charging input.

Question: #102
Answer D is correct. Selling a part with a core requires that the core be included on the invoice and billed to the customer. Once the core is returned the customer will receive credit. The only way to track the return of cores is to invoice the customer and match the core to the billed invoice. Cores are very valuable and are a source of lost revenue within a store operation. Successfully accounting for cores will enhance the businesses bottom line.

Answers to the Test Questions for the Additional Test Questions Section 6

1.	A	28.	C	55.	C	82.	B
2.	D	29.	C	56.	C	83.	D
3.	D	30.	B	57.	B	84.	A
4.	A	31.	D	58.	D	85.	C
5.	C	32.	A	59.	C	86.	D
6.	A	33.	A	60.	A	87.	B
7.	A	34.	C	61.	C	88.	A
8.	A	35.	A	62.	B	89.	B
9.	B	36.	A	63.	A	90.	C
10.	A	37.	D	64.	C	91.	B
11.	C	38.	D	65.	C	92.	D
12.	B	39.	C	66.	B	93.	C
13.	B	40.	B	67.	B	94.	D
14.	D	41.	B	68.	A	95.	B
15.	C	42.	C	69.	C	96.	C
16.	A	43.	C	70.	A	97.	D
17.	A	44.	B	71.	C	98.	C
18.	C	45.	D	72.	A	99.	D
19.	B	46.	B	73.	B	100.	D
20.	A	47.	C	74.	D	101.	D
21.	D	48.	B	75.	C	102.	A
22.	C	49.	A	76.	C	103.	B
23.	A	50.	D	77.	C	104.	C
24.	C	51.	A	78.	C	105.	B
25.	A	52.	C	79.	B		
26.	D	53.	D	80.	D		
27.	C	54.	C	81.	D		

Explanations to the Answers for the Additional Test Questions Section 6

Question: #1
Answer A is correct. Parts Specialist A is correct; a 20 percent discount on a $60.00 part would be $12.00. To calculate the percent discount, convert the percent in a decimal number by moving the decimal two places to the left. For example, 20 percent becomes 0.20, then multiply the decimal number by the original price to get the amount of the discount. As in the question, 20 percent of $60.00 is $12.00 (0.20 × 60.00 = 12.00) Parts Specialist B is incorrect. To calculate the percent of gross margin profit, subtract the cost from the selling price to determine the gross profit dollars, then divide the gross profit dollars by the selling price to calculate the percentage of gross profit. ($175.00 – $100.00 = $75.00) then divide ($75.00 ÷ $175.00 = 42.9%).

Question: #2
Answer D is correct. The restock fee, usually expressed as a percentage, is the fee charged for having to handle a returned part. To calculate the restocking fee, convert the percent into a decimal number by moving the decimal two places to the left. 10% becomes 0.10. Multiply the decimal number by the original price to get the amount of the restocking fee. 0.10 × $250.00 is $25.00. Subtract the restocking fee ($25.00) from the original price ($250.00) to get the amount of money to be returned to the customer. $250.00 – $25.00 = $225.00.

Question: #3
Answer D is correct. Neither parts specialist is correct. There are 61 cubic inches to a liter. To convert to cubic inches, multiply the liters by 61 to get cubic inches. For example, a 5-liter engine is the same as a 305 cubic inch engine (5 × 61 = 305). The metric unit of temperature measurement is degrees Celsius (° C). To convert from degrees Fahrenheit (° F) to degrees Celsius using a short formula, first subtract 32 then multiply by 0.555. For example 212° F is equal to 99.9° C (212 – 32 = 180, and 180 × 0.555 = 99.9).

Question: #4
Answer A is correct. An alphanumeric listing places a series of numbers and letters in order starting from the left digit and working across to the right. If the numbers are the same, the order sequence continues with the letter A, then B, and so forth.

Question: #5
Answer C is correct. Reading a micrometer is done by approaching the measurement in steps. Looking at the figure in this question you can see the measurement is 0.245. Because this is a 0–1 inch micrometer, the reading must be between 0 and 1. Since the 2 is the last complete unit visible on the horizontal line, the first number is 0.200. One $\frac{1}{25}$-inch mark is visible after the 2, so the second measurement is 1 × 0.025 = 0.025. The horizontal line nearly lines up with the 20 on the vertical line, so 0.020 is the third measurement. Adding up the three measurements (0.200 + 0.025 + 0.020) gives a total reading of 0.245.

Question: #6
Answer A is correct. It is the responsibility of all employees to keep all areas of the store clean and orderly. All employees suffer from lost sales if the floors, shelves, and displays are a safety hazard or do not appeal to the customer.

Question: #7
Answer A is correct. The Environmental Protection Agency (EPA), Occupational Safety and Health Administration (OSHA), and other state and local agencies have strict guidelines for handling these materials. OSHA's Hazard Communication Standard (HCS), commonly called the "Right-to-Know Law," applies to all companies that use or store any kind of hazardous chemicals that workers might come in contact with, including solvents, caustic cleaning compounds, abrasives, cutting oils, and other hazardous materials. Compliance with the law requires a system for labeling hazardous chemicals and maintaining Material Safety Data Sheets (MSDS). These sheets must be made available to employees, informing them of the dangers inherent in chemicals found in the workplace.

Question: #8
Answer A is correct. Specialist A is correct. This setup is used to see how straight the rotor is. Any deflection shown on the dial indicator shows how un-flat the rotor is. Rotor thickness is measured with a micrometer.

Question: #9
Answer B is correct. Expensive items should be displayed behind the counter in locked cases. Parts specialists should give a lot of thought to what gets displayed near entrances and exits. In the moment that a parts specialist is busy researching parts in a catalog, a thief can easily slip something under his or her jacket and glide through the door. It is best to keep the areas around the entrances clear or to display only large, heavy, or awkward merchandise near the door. Barrels of oil and bags of floor sweep are safe choices for displays near entrances.

Question: #10
Answer A is correct. Do-it-yourselfers usually need more information than do professional automotive customers. These customers usually need and expect good advice on parts and methods. They view the parts specialist as an expert, and the parts specialist should utilize product knowledge and catalog skills to give the advice needed. Nothing should be sold to a customer that is not actually needed. One way that a parts specialist can assist do-it-yourselfers without taking too much time away from other customers is to have printed how-to information available. Once the parts specialist determines what work the customer will be doing, the parts specialist can provide something to read while assisting other customers. After the customer reads the information, the parts specialist can clarify some important points and answer questions. Each customer's needs can be met without anyone waiting a long time. Since your job is to assist customers, you should take the time that is needed to help them. However, you should do what you can to control that time.

Question: #11
Answer C is correct. Both parts specialists are right. A parts specialist's opening question should be designed to put the customer at ease and elicit as much data as possible. The question, "May I help you?" is the worst possible opening because it invites the opportunity for the customer to reply, "No." Instead, the parts specialist should ask, "How can I help you?" Another good opening is, "Can I show you anything in particular, or would you like to browse around first?" These questions restrict answers to the positive and give the customer a way out without having to say, "No." The customer's response will generally indicate whether he or she has a defined objective, is just looking, or is seriously considering a purchase.

Question: #12
Answer B is correct. The tool being used is a drum brake tool designed to match the diameter of the drum to the expanse of the shoes. This procedure is used to set the shoes prior to installing the drum. This sets the shoes so that they need a minimum amount of adjustment.

Question: #13
Answer B is correct. Often the phone will ring while the parts specialist is serving a customer. If no one else is available to answer the phone, the parts specialist should politely ask the customer to wait and then answer the phone. If it appears that the call will take a while, the parts specialist should explain to the caller that he or she is assisting another customer at the moment. After taking pertinent information, the parts specialist should state that he or she will return the call as soon as possible or that the caller should phone back in a few minutes. That way, neither the caller nor the counter customer is kept waiting very long. Since a caller cannot see the parts specialist's facial expressions, gestures, or body language, all communications must be accomplished with words, tones, timing, and inflection. Parts counter personnel must remember to speak clearly and slowly enough to be understood and must request that the customer do so if not clearly understood.

Question: #14
Answer D is correct. The appearance of the store, as well as the appearance of counter personnel, influences the customer. A clean, neat parts specialist establishes a positive image with customers, and a clean, neat store is equally important. Although a parts specialist is rarely in charge of decor or layout, he or she can maintain the appearance of the store in several ways: keep a rag handy for wiping the counter so that it is free of dirt and grime from used parts; keep the counter free of clutter such as small parts, notes, paper clips, and other such items; keep the displays organized and well stocked; keep the stock and supplies neatly shelved to improve appearance; and assist with basic clean-up by throwing away empty parts wrappers and labels and by removing empty boxes from sight.

Question: #15
Answer C is correct. Both parts specialists are right and If the parts specialist suggests related items on the mechanic's first visit, both the customer and the parts specialist can save a lot of time. The customer will be very appreciative of the parts specialist's initial time investment. With this type of service the customer will more than likely return for parts again. Selling related parts when a customer buys a particular item can boost profits by 30 percent or more. Selling related parts also makes sense when the parts specialist remembers that a customer's problem might not be solved solely by replacing a faulty part if the related hardware or chemicals are not also up to peak performance. This is especially important if a related part is on sale.

Question: #16
Answer A is correct. The procedure shows using plastigage to check bearing clearance. If the results indicate that there is too little or too great a clearance, the bearing may need to be replaced with one of a different thickness. Shaft out-of-roundness would probably not affect the results of this check.

Question: #17
Answer A is correct; the part shown is a typical in-line fuel filter.

Question: #18
Answer C is correct. The procedure shown is for installing new threaded inserts into bores that had damaged threads. These inserts are available in most thread pitches and sizes.

Question: #19
Answer B is correct. The procedure shown is the typical way to adjust valve lash or clearance on an OHV engine. The lash is measured with a feeler gauge and is adjusted with the adjustment screw. The lock nut is used to hold that adjustment.

Question: #20
Answer A is correct. The tester shown in the figure is a standard cooling system tester. It is used to check for leaks throughout the system, including the radiator cap. The tester is used to create a high pressure in the system or on the radiator cap. If there is a leak, the pressure will not build up. In the case of a radiator cap, pressure should be able to build up to the pressure rating of the cap. If the cap cannot hold a lower pressure, it is bad and should be replaced. The tester does not create a vacuum. Therefore, choice A is wrong, but for this question—that is the correct answer.

Question: #21
Answer D is correct. If you understand the major components of an A/C system and look carefully at the arrangement of the parts, you should be able to identify the parts. Also, as a hint, the drawings give you an outline of the shape of the component. In the drawing, component 1 is the compressor, 2 is the condenser, 3 is the evaporator, 4 is the accumulator, and 5 is a pressure switch. A radiator and air pump are not part of an A/C system.

Question: #22
Answer C is correct. Both parts specialists are correct. In most receiver drier assemblies, the desiccant bag is replaceable without invading the A/C system. Eventually these bags become saturated and can no longer remove moisture from the system. Because this unit is a VIR, it controls refrigerant flow to the evaporator as well as serves as a receiver/drier.

Question: #23
Answer A is correct. The part shown is an ignition coil for a distributorless ignition system.

Question: #24
Answer C is correct. The thread pitch of a bolt in the fractional system is determined by the number of threads there are in one inch of threaded bolt length and is expressed in number of threads per inch. The thread pitch in the metric system is determined by the distance in millimeters between two adjacent threads. To check the thread pitch of a bolt or stud, a thread pitch gauge is used. Gauges are available in both English and metric dimensions.

Question: #25
Answer A is correct. Only Parts Specialist A is right. The standard location for the vehicle identification number (VIN) is attached to the driver side of the instrument panel and is visible through the windshield.

Question: #26
Answer D is correct because both parts specialists are wrong. The standard location for the vehicle identification number (VIN) is attached to the driver side (not passenger side) of the instrument panel and is visible through the windshield. The production date is the date that the vehicle was assembled. The standard location for this information is on a tag affixed to the driver's door by the door latch or on the driver's door sill plate. It certainly would not be found on the spare tire cover.

Question: #27
Answer C is correct. Both specialists are correct. Component identification (ID) data are stamped in the casting or on a tag that is similar to the one shown in the figure and attached to the component. All major components on the vehicle will have ID data attached.

Question: #28
Answer C is correct. Both parts specialists are right. The paint code is not part of the vehicle identification number (VIN). It is located on a separate trim tag, called the service parts identification label, which is mounted under the hood or on the driver's door of some vehicles.

Question: #29
Answer C is correct because both parts specialists are right. Each catalog has, in the upper right-hand corner, the form number, the date of issue, and the number and date of the catalog it replaces or supplements. The cover also identifies the product line and manufacturer and often provides an area where the jobber can stamp his or her name for the convenience of dealer customers. If catalogs were not the same size, they would be difficult to organize and would make the job of locating parts very difficult.

132   Explanations to the Answers for the Additional Test Questions Section 6 — Appendices

Question: #30
Answer B is correct. A footnote is used for referencing additional information. Every parts specialist will have occasion to refer to footnotes in the catalogs at some point, either to reference additional information, determine exceptions, or find alternative parts. Sometimes footnotes are difficult to understand when a number of them are squeezed on a page. The footnotes themselves are important. Footnotes should be referenced systematically in order to fully define those applications that have multiple possibilities.

Question: #31
Answer D is correct. Correction bulletins are used to make corrections during catalog maintenance. Correction bulletins refer to catalog errors due to mistyping or inaccurately assigned parts numbers.

Question: #32
Answer A is correct. The abbreviation DLC means Data Link Connector.

Question: #33
Answer A is correct. Parts Specialist A is right. A seasonal item is something that a parts department may not normally stock during certain times of the year. Some seasonal items are stocked throughout the entire year but not in large quantities. An example of one of these items is windshield washer fluid, which is stocked more heavily in the winter season than the summer season but is still sold throughout the year. Another such example is car batteries, which are stocked all year even though battery failures occur more often during the colder parts of the year.

Question: #34
Answer C is correct. In order to keep shelves and items stocked and looking appealing to a customer, the shelves must be well maintained. The shelves, as well as the items on the shelves, should not be dirty or dusty. Dusty items will not sell well because nobody wants to buy anything that they may feel is old and has been sitting around forever. The items on the shelves should be replenished on a regular basis to encourage repeat customers to count on you to have the items they need at all times. If shelves are messy and items are not easily found, this will discourage a sale or a customer from revisiting your store or department again. A shelf should be labeled to some extent to let people know which items go where. Displaying a price on a shelf where the item goes will help keep shelves in order because a customer will not have to take the item off the shelf to see how the price may differ from a similar item next to it. A customer may not always return the item to the space where it was removed, causing an unorganized appearance.

Question: #35
Answer A is correct. A normal practice of display pricing is to display an item that may be specially priced for that week in an eye-catching display in an open part of the store or department, along with a noticeable sign that displays the special pricing. The product that is specially priced may still remain in its assigned shelf position, but this additional display should let customers know this item is on special. A customer is more likely to notice a display with a posted price rather than notice an item on the shelf in its normal position. This kind of display and pricing is also used for the introduction of new items. A vendor may also choose to put up a display if overstocked in a certain item but choose not to lower the price. This will help promote the item and possibly lead to more sales. Display pricing can also be used to bring customers to the department or store by the distribution of flyers with an explanation and the price, by advertising on a billboard with the price, or signs in a window with the price.

Appendices
Explanations to the Answers for the Additional Test Questions Section 6 133

Question: #36
Answer A is correct. The best place for a display is in an open area near the front of the store or department. This ensures that the display will catch the customer's eye as soon as he or she enters. The display should be large and noticeable, and it must look appealing to the customer. This is usually accomplished by building the display out of the product being sold. Often the manufacturer of a product will send a display kit that can be assembled in the store and which may include banners, signs, and shelves. Displays are also often built at the end of an isle where the product is located so the customer will notice the product without entering the isle. Displays can be set up with the intent of the customer taking the actual item from the display. The display should be set up with care to keep it from falling apart as items are removed. A display also needs to be maintained for a neat appearance and to keep it appealing to the customer.

Question: #37
Answer D is correct. A customer orders six liters of transmission fluid. You should give the customer 6 quarts of ATF. One quart is equal to 0.9464 liters, so one quart is almost equal to one liter. Therefore, 6 liters is approximately equal to 6 quarts. In actual practice, this amount will leave the transmission about ¼ quart low on fluid but within the safe operating range.

Question: #38
Answer D is correct. Looking at the invoice in the figure, you can see that is true. Answer A is cost of dressing and choice B is the customer's cost or sales price for the dressing. Answer C is the cost of the belt but not what the customer paid.

Question: #39
Answer C is correct because both parts specialists are right. Charge account sales encourage large purchases or purchases from customers who do not want to carry cash or do not have the cash. Counter personnel must be familiar with the charge policy in their workplace to ensure accuracy and customer satisfaction.

Question: #40
Answer B is correct. Parts Specialist A is wrong. A parts specialist should seek whatever help is necessary in order to sell the correct parts and satisfy the customer. Specialist B is right; experienced parts specialists are often asked to help train new employees.

Question: #41
Answer B is correct. The scale on the micrometer shown in the figure shows that the micrometer is an inch-type. Therefore Parts Specialist A is wrong in saying the reading in the figure is 0.560 mm. The numbers were right but the measurement wrong. Parts Specialist B is correct in saying the reading is 0.560 inches.

Question: #42
Answer C is correct. It is the responsibility of all employees to keep all areas of the store clean and orderly. All employees suffer from lost sales if the floors, shelves, and displays are a safety hazard or do not appeal to the customer. A parts specialist is not responsible for keeping the shop parts clean or tidy.

Question: #43
Answer C is correct because both parts specialists are right. Small parts usually do not weigh much; therefore, there is no need to seek help when handling them. Knowing the proper way to lift and work within one's ability is important for handling parts. Shipping, receiving, and stocking materials require physical exertion. Knowing the proper way to lift heavy materials is important. Always lift and work within your ability and seek help from others when you are not sure if you can handle the size or weight of the material or object. Auto parts, even small, compact components, can be surprisingly heavy or unbalanced. Always size up the lifting task before beginning.

Question: #44
Answer B is correct. The Occupational Safety and Health Administration (OSHA) is an organization that sets the guidelines for companies that use hazardous chemicals.

Question: #45
Answer D is correct. Expensive items should be displayed behind the counter in locked cases. Parts specialists should give a lot of thought to what gets displayed near entrances and exits. In the moment that a parts specialist is busy researching parts in a catalog, a thief can easily slip something under his or her jacket and glide through the door. It is best to keep the areas around the entrances clear, or to display only large, heavy, or awkward merchandise near the door. Barrels of oil and bags of floor sweep are safe choices for displays near entrances.

Question: #46
Answer B is correct. Parts Specialist A is wrong; a parts specialist should not sell the customers what they want if it will not fix the problem. Parts Specialist B is correct in saying only the parts that are needed for a particular repair should be sold to the customers. Do-it-yourselfers usually need more information than do professional automotive customers. These customers usually need and expect good advice on parts and methods. They view the parts specialist as an expert, and he or she should utilize product knowledge and catalog skills to give the advice needed. Nothing should be sold to a customer that is not actually needed.

Question: #47
Answer C is correct. Both specialists are right. A telephone customer cannot see the parts specialist's body language so all communication must be accomplished with words, tones, triming, and inflection. Parts specialists must remember to speak clearly and slowly enough to be understood and must request that the customer do so if he or she cannot be understood.

Question: #48
Answer B is correct. The part shown is a ball joint. A ball joint allows the control arms to move up and down and the spindle to rotate (right and left) when the steering wheel is turned.

Question: #49
Answer A is correct. Although a parts specialist is rarely in charge of decor or layout, he or she can maintain the appearance of the store in several ways: keeping a rag handy for wiping the counter so that it is free of dirt and grime from used parts; keeping the counter free of clutter such as small parts, notes, paper clips, and other such items; keeping the displays organized and well stocked; keeping the stock and supplies neatly shelved to improve appearance; and assisting with basic clean-up by throwing away empty parts wrappers and labels and by removing empty boxes from sight. Parts Specialist A is right.

Question: #50
Answer D is correct. The procedure is called bench bleeding a master cylinder. This procedure is recommended prior to installation of the unit to a vehicle. Bench bleeding removes the air from the internal parts of the master cylinder. Failure to do this will make bleeding of the brakes, after installation, very difficult. Fluid is put in the master cylinder's reservoir and the dowel is used to move the pistons in the cylinder. This movement pumps the fluid through the cylinder and pushes the air out. The tubes and dowel must be removed prior to installation.

Question: #51
Answer A is correct. Parts Specialist A is right. Front-wheel-drive vehicles have a driveshaft for each front wheel. Parts Specialist B is wrong. The inboard constant velocity (CV) joint is typically a plunger type.

Question: #52
Answer C is correct. Specialist A is right. The part is an outboard CV joint. It is being removed from an axle or half shaft. The retaining ring holds the joint to the shaft. Replacement of this type of retainer is always recommended when they are removed. When they are expanded for removal, they lose some of their spring tension. To prevent failure, they should be replaced. Both specialists are right.

Question: #53
Answer D is correct. Both specialists are wrong. In both cases the spring should be replaced. Heating the spring will cause it to lose tension rendering it near useless. Shims are placed under the spring to set the tension at a specified installed height. If the spring is shorter than it should be, it has collapsed and needs to be replaced.

Question: #54
Answer C is correct because both parts specialists are right. A vehicle's electrical system is protected by fuses, circuit breakers, and fusible links. These protection devices will blow or open if there is a shorted circuit.

Question: #55
Answer C is correct. In the fractional system, the tensile strength of a bolt is identified by the number of radial lines (grade marks) on the bolt head. More lines mean higher tensile strength. In the metric system, tensile strength of a bolt or stud can be identified by a property class number on the bolt head. The higher the number, the greater the tensile strength. A source for confusion in the markings is best demonstrated by looking at a bolt with no (zero) grade marks on the bolt head. This bolt would be a grade 0 bolt. If the bolt had three grade marks (lines) on its head, it would be a grade 5 bolt.

Question: #56
Answer C is correct. Both parts specialists are right. Component identification (ID) data are stamped in the casting or on a tag that is attached to the component. All major components on the vehicle will have ID data attached.

Question: #57
Answer B is correct. The pickup truck shown in the figure for this question has a regular cab, one that has one seating surface and two doors. Therefore Specialist A is wrong. Specialist B is correct in saying that an extended cab vehicle could have two, three, or four doors.

Question: #58
Answer D is correct. Correction forms are not included in most catalogs. Most catalogs contain form number(s), date of issue, and the name of product lines and manufacturers.

Question: #59
Answer C is correct. Both parts specialists are right. A supersession bulletin may include part numbers that supersede previously issued part numbers and technical bulletins may alert counter personnel to any unusual installation or application problems.

Question: #60
Answer A is correct. Parts Specialist A is right in saying that to verify an abbreviation one should refer to the abbreviation list. Parts Specialist B is wrong. Manufacturers usually include additional aids for using their publications such as an abbreviation list, which can make for efficient use of the catalog. For example, abbreviations can often have more than one meaning. FWD can mean front-wheel drive or four-wheel drive. OD can mean overdrive or outside diameter. The definitions of these, and other abbreviations, can only be determined by checking the abbreviation list provided.

Question: #61
Answer C is correct. A seasonal item is something that a parts department may not normally stock during certain times of the year. Some seasonal items are stocked throughout the entire year but not in large quantities. An example of one of these items is windshield washer fluid, which is stocked more heavily in the winter season than the summer season but is still sold throughout the year. Another such example is car batteries, which are stocked all year even though battery failures occur more often during the colder parts of the year. Some other items used year-round but stocked more heavily in the winter include antifreeze, winter windshield wipers, and gasoline additives to prevent gas line freeze. Some items more heavily stocked in the summer include air conditioning refrigerant, windshield sunshades, and wax or polish. An impulse item is a product that the customer buys on the spur of the moment to fill a "want" rather than a "need" for the item. Some examples of impulse items include vehicle appearance-enhancing items such as pin striping kits, chrome accessories, high-flow air filters to improve vehicle performance, and interior items like extra cup holders or seat covers.

Question: #62
Answer B is correct. Returned items should not be put back on the shelf if they are opened as this will give them a used appearance and make the customer uneasy about purchasing them. Items that are returned unopened can be returned to the shelf if they appear to be in good order. In order to keep shelves and items stocked and looking appealing to a customer the shelves must be well maintained. The shelves, as well as the items on the shelves, should not be dirty or dusty. Dusty items will not sell well because nobody wants to buy anything that they may feel is old and has been sitting around forever. The items on the shelves should be replenished on a regular basis to encourage repeat customers to count on you to have the items they need at all times. If shelves are messy and items are not easily found, or shelves are always empty from not restocking, often enough this will discourage a sale or a customer from revisiting your store or department again. The next item on a shelf should be pulled forward when the one in front of it is sold. This will give your shelf the appearance of being full and items will be easier to find for the next customer. When items are restocked the new items should be put behind the old ones. This will help keep dust from building up on items. A shelf should be labeled to some extent to let people know which items go where. Displaying a price on a shelf where the item goes will help keep shelves in order because a customer will not have to take the item off the shelf to see how the price may differ from a similar item next to it. A customer may not always return the item to the space where it was removed, causing an unorganized appearance.

Question: #63
Answer A is correct. A normal practice of display pricing is to display an item that may be specially priced for that week in an eye-catching display in an open part of the store or department, along with a noticeable sign that displays the special pricing. The product that is specially priced may still remain in its assigned shelf position, but this additional display should let customers know this item is on special. A customer is more likely to notice a display with a posted price rather than notice an item on the shelf in its normal position. This kind of display and pricing is also used for the introduction of new items. A vendor may also choose to put up a display if overstocked in a certain item but choose not to lower the price. This will help promote the item and possibly lead to more sales. Display pricing can also be used to bring customers to the department or store by the distribution of flyers with an explanation and the price, by advertising on a billboard with the price, or signs in a window with the price.

Question: #64
Answer C is correct. The LEAST important part of a display is the product on the display. The best place for a display is in an open area near the front of the store or department. This ensures that the display will catch the customer's eye as soon as he or she enters. The display should be large and noticeable and it must look appealing to the customer. This is usually accomplished by building the display out of the product being sold. Often the manufacturer of a product will send a display kit that can be assembled in the store and which may include banners, signs, and shelves. Displays are also often built at the end of an isle where the product is located so the customer will notice the product without entering the isle. Displays can be set up with the intent of the customer taking the actual item from the display. The display should be set up with care to keep it from falling apart as items are removed. A display also needs to be maintained for a neat appearance and to keep it appealing to the customer.

Appendices   Explanations to the Answers for the Additional Test Questions Section 6   137

Question: #65
Answer C is correct. An impulse item is a product that the customer buys on the spur of the moment to fill a "want" rather than a "need" for the item. Some examples of impulse items include vehicle appearance-enhancing items such as pin striping kits, chrome accessories, high-flow air filters to improve vehicle performance, and interior items such as extra cup holders or seat covers.

Question #66
Answer B is correct. Parts Specialist A is wrong; 327 cubic inches is equal to about 5.2 liters. Parts Specialist B says that 488 cubic inches is equal to 8 liters. There are 61 cubic inches to a liter.

Question: #67
Answer B is correct. Parts Specialist A is wrong. The markup is 14 cents. Sixty-one cents is the difference between the cost of one nut and the sale price of two. Parts Specialist B is right; the markup on the camshaft is 40 percent. To determine the markup, divide the actual amount of markup (in dollars) by the cost. In this case, the actual markup is $124.85 – 89.18, which equals $35.67. Now divide that amount by $89.18. The answer is 40 percent.

Question: #68
Answer A is correct. An invoice should be completed for each sale in the store. These invoices are used to track inventory and customer purchases for billing purposes. A purchase order is used to allow a company to purchase parts. It will describe the parts and quantity of parts to be purchased along with billing information. A stock order is used by the store to order more stock from the suppliers. A back order refers to merchandise ordered from a supplier but not shipped, due to the supplier being out of stock. An emergency order is an order placed with the supplier on a routine basis, usually weekly or biweekly.

Question: #69
Answer C is correct, because both parts specialists are right. When the parts specialist carries many different brands of the same product, the difference between them should be explained. If the store carries, for example, four different kinds of brake pads, the parts specialist should identify the material, size, cost, warranty, and name brand differences. The parts specialist should let the customer decide on which part to buy. The parts specialist can make recommendations, but the final decision should be the responsibility of the customer.

Question: #70
Answer A is correct. If a parts specialist must put a telephone customer on hold, he or she should identify the name of the store and politely state something such as, "Please hold one moment." The parts specialist should never pick up the phone and press the hold button without saying anything, and he or she should never push the button before finishing a sentence or phrase. If the parts specialist knows that it will take a while before he or she can talk with the caller, the parts specialist should ask that the caller call back rather than be put on hold for a long period of time.

Question: #71
Answer C is correct. Selling related items is extremely important, not only to build profits, but also to help the customer achieve the safest possible results. Displays should be set up to advertise the complete service job. Brake pads, shoes, and hardware once came in fairly plain, dull boxes, but today, most manufacturers have upgraded their packaging graphics to the point where these products can be used to build attractive, effective displays in the front of the store. Both specialists are right and

Question: #72
Answer A is correct. Parts Specialist A is right; product additions can be found in the table of contents of any parts catalog. Parts Specialist B is wrong. The table of contents and index will show the catalog sequence. It will be quite evident if trucks are listed with passenger cars, or if imports are integrated or listed separately. Product additions, possibly resulting from the consolidation of one or more formerly separate catalogs, can be quickly spotted. Instant recognition of several other catalog entities can also be found. These include numerical listings, buyer's guides, progressive size listings, illustrations, installation instructions, competitive interchanges, and more.

Question: #73
Answer B is correct. Parts Specialist A is wrong, but Specialist B is right. The table of contents and index will show the catalog sequence. It will be quite evident if trucks are listed with passenger cars, or if imports are integrated or listed separately. Product additions, possibly resulting from the consolidation of one or more formerly separate catalogs, can be quickly spotted. Instant recognition of several other catalog entities can also be found. These include numerical listings, buyer's guides, progressive size listings, illustrations, installation instructions, competitive interchanges, and more.

Question: #74
Answer D is correct. As with any delivery, an order for a shipment must be carefully picked, packaged, and documented. A packing list or slip must be included with each shipment. This list contains a detailed description of the items included in the shipment. A separate packing list can be placed in each part of a multiple-part shipment, but often a single comprehensive list is packaged with or secured to the first part of a multiple-part shipment. The only correct statement of the answer choices is D.

Question: #75
Answer C is correct. A packing list or slip must be included with each shipment. This list contains a detailed description of the items included in the shipment. A separate packing list can be placed in each part of a multiple-part shipment, but often a single comprehensive list is packaged with or secured to the first part of a multiple-part shipment. Both specialists are right.

Question: #76
Answer C is correct. Physical inventory should be done once a year. Physical inventory is when all stock is pulled off of the shelves and checked to determine what is actually in the store or shop versus what is in the computer. This is one way to keep the computer system up-to-date and can help you to give the customer faster and more accurate service. Both parts specialists are right and

Question: #77
Answer C is correct. After the physical inventory has been completed, the parts specialist should report all discrepancies. The discrepancies are reported by filling out an inventory discrepancy form. After discrepancies have been entered into the computer system, they should be kept and filed for future reference.

Question: #78
Answer C is correct. Parts Specialist A is right. All of the discrepancy forms should be kept and filed. Parts Specialist B is also right. After a discrepancy form has been filled out, it must be corrected in the computer system.

Question: #79
Answer B is correct. Stock should be rotated every time the shelves are stocked. Stock rotation is moving the older stock in front of the newer stock. This prevents the older stock from sitting at the back of the shelf and eventually getting an aged appearance. Rotating the product so that the labels are toward the front of the shelf is known as facing. Rearranging the shelves for a neater appearance is done whenever necessary to increase appeal to the customer.

Question: #80
Answer D is correct. A core charge is a charge that is added when the customer buys a remanufactured or reconditioned part. Core chargers are refunded to the customer when the old defective but rebuildable part is returned. To ensure that customers return their cores, a parts specialist should always require a core charge. In this question all of the parts listed may be available for remanufacture, except spark plugs. Often brake shoes, brake calipers, and water pumps are sold as rebuilt units.

Question: #81
Answer D is correct. A warranty-return part should be kept off the shelf, not to be sold again. Additional forms must usually be completed for warranty-return parts. This allows the manufacturers to determine what is wrong with the part and to credit the store for the return. A warranty-return part should never be placed back in the inventory to be sold to another customer.

Question: #82
Answer B is correct. A pair refers to 2.

Question: #83
Answer D is correct. Each refers to one (1).

Question: #84
Answer A is correct. Parts Specialist A is right and B is wrong. Freight charges may be added to emergency order parts to cover the cost of special transportation to the store. It is not uncommon to add a long-distance phone charge for special phone orders. Some parts may have a restocking fee that is charged to the customer when the parts are returned for a refund.

Question: #85
Answer C is correct. The main benefit of using a lost sales report is to keep track of parts that are selling and those that are not selling. Using lost sales reports allow you to spot trends in the aftermarket parts business. These reports will show new part numbers that the store does not stock as well as older part numbers that are increasingly in demand. Some part numbers are assured of almost instant demand, such as those found in some of the newer downside models. Others parts will grow in demand, but somewhat more slowly. The storeowner or buyer must analyze not only the part numbers that are selling, but also those that the store does not stock.

Question: #86
Answer D is correct. The type of report shown in the figure is a lost sales report.

Question: #87
Answer B is correct. Parts Specialist A says the positive crankcase ventilation (PCV) valve clean air filter in the air cleaner may be plugged. If this were the case, oil that would enter from the breather hose would not be able to enter the air cleaner housing. Therefore A is wrong. Parts Specialist B says the hose from the positive crankcase ventilation (PCV) valve to the intake manifold may be severely restricted. This would cause crankcase pressures to escape through the PCV air filter and collect in the air cleaner housing.

Question: #88
Answer A is correct. Repeated extension housing seal failure may be caused by a scored driveshaft yoke. The other answer choices would have little to no effect on the wear of the seal.

Question: #89
Answer B is correct. Automatic transmission fluid (ATF) is pink or red. If the fluid is dark brown or blackish and has a burned odor, the fluid has been overheated, possibly from burned clutches or bands. A milky-colored fluid is Most-Likely caused by coolant contamination from a leaking transmission cooler. If silvery metal particles are found in the fluid, it is an indication of damaged transmission components. If the dipstick feels sticky and is difficult to wipe clean, the fluid contains varnish, an indication that ATF and filter changes have been neglected. In this question the fluid is a dark brown color and has a burned smell. Parts Specialist A is wrong. This problem may be caused by burned clutches or bands, not by a worn front planetary sungear.

Question: #90
Answer C is correct. A worn U-joint may cause a squeaking noise that decreases in relation to vehicle speed. This happens more in vehicles that have higher driveline angles like 4WD trucks. Commonly a worn joint will cause a clanking noise when the transmission shifts gear or when the vehicle is accelerating or decelerating. Parts A heavy vibration that only occurs during acceleration may be caused by a worn centering ball and socket on a double cardan U-joint.

Question: #91
Answer B is correct. The item in the drawing is a thermal expansion valve.

140   Explanations to the Answers for the Additional Test Questions Section 6   Appendices

Question: #92
Answer D is correct. Both parts specialists are wrong. When measuring resistance with an ohmmeter, the circuit or component must be disconnected from power. If this is not done, the circuit's current will damage the meter. Also, when testing a spark plug wire with a 20,000 Ohm resistance, you should use the ×1K scale not the ×100 scale.

Question: #93
Answer C is correct. Both specialists are right. Gasoline engines and diesel engines are internal combustion engines.

Question: #94
Answer D is correct. Flywheel run out is measured with the use of a dial indicator. Inspect the flywheel for scoring and cracks in the clutch contact area. Minor score marks and ridges may be removed by resurfacing the flywheel. Mount a dial indicator on the engine flywheel housing, and position the dial indicator stem against the clutch contact area on the flywheel. Rotate the flywheel to measure the flywheel run out. If the flywheel run out exceeds specifications it must be replaced.

Question: #95
Answer B is correct. Blue exhaust is an indication of excessive oil consumption. A sound engine would be nearly invisible. The exhaust from an engine that has a rich air/fuel mixture would have a blackish exhaust. Coolant leaking into the combustion chamber would cause a whitish exhaust.

Question: #96
Answer C is correct. A missing air filter and a clogged fuel filter can affect engine performance.

Question: #97
Answer D is correct. Worn turbocharger seals may be noted by blue smoke from the exhaust, due to the oil in the exhaust.

Question: #98
Answer C is correct. The statement that is not true is C. The meter is connected to measure the resistance across a diode. A good diode will have high resistance when the meter is connected in one direction. If the leads of the meter are reversed, the meter should show a low resistance reading.

Question: #99
Answer D is correct. A typical exhaust system has the following components: exhaust manifold and gasket; exhaust pipe, seal, and connector pipe; intermediate pipe(s); catalytic converter; resonator and/or muffler; tailpipe; and hardware items including heat shields, clamps, gaskets, and hangers. Of the answer choices, D is the item not found on all vehicles.

Question: #100
Answer D is correct. The evaporative emission control system draws vapors from the fuel tank and introduces them into the intake air stream.

Question #101
Answer D is correct. The drive shaft is the connection between the transmission output shaft and the differential. Composite drive shafts are used in some high performance vehicles. Titanium is not currently used to produce original equipment drive shafts.

Question: #102
Answer A is correct. Parts Specialist A is right. Constant velocity (CV) joints are serviced as a complete assembly. Parts Specialist B is wrong. When reassembling the joint always install all the grease in the joint that is provided in the repair kit.

Question: #103
Answer B is correct. Specialist B is correct. This setup is used to measure the thickness of a rotor. This thickness is compared to the minimum thickness specs. If the rotor is less than specifications, the rotor should be discarded. Rotor distortion is checked with a dial indicator.

Question: #104
Answer C is correct. Both parts specialists are right. Bolt diameter is the measurement across the major diameter of the threaded area or across the bolt shank. The thread pitch of a bolt in the English system is determined by the number of threads there are in one inch of threaded bolt length and is expressed in number of threads per inch. The thread pitch in the metric system is determined by the distance in millimeters between two adjacent threads. To check the thread pitch of a bolt or stud, a thread pitch gauge is used. Gauges are available in both English and metric dimensions. Bolt length is the distance measured from the bottom of the head to the tip of the bolt. The bolt's tensile strength, or grade, is the amount of stress or stretch it is able to withstand. The type of bolt material and the diameter of the bolt determine its tensile strength. In the English system, the tensile strength of a bolt is identified by the number of radial lines (grade marks) on the bolt head. More lines mean higher tensile strength. In the metric system, tensile strength of a bolt or stud can be identified by a property class number on the bolt head. The higher the number, the greater the tensile strength.

Question: #105
Answer B is correct. The tool shown is plastigage and is used to measure the clearance between the bearing and shaft. A strip of thread is placed between the shaft and the bearing. How much the thread spreads depends on the clearance. The less clearance, the more the thread will spread.

Glossary

Accessories Parts that add to the appearance or performance of a vehicle.

Accounts receivable Money due from customers.

Accumulator Part of an air conditioning system which contains a desiccant that absorbs moisture from the refrigerant.

Active accounts Current customers who make frequent purchases.

Active stock Merchandise in the store that is readily available for sale to customers.

Air filter Part of an engine intake system which cleans dirt and dust from the air before it enters the engine.

Air-Injection Reaction (AIR) An emissions control system which pumps air into the exhaust system to burn hydrocarbons, carbon monoxide, and oxides of nitrogen coming from the engine.

Air pump Part of the Air Injection Reaction (AIR) system which forces extra air to mix with the exhaust gases. This action causes the continued burning of any hydrocarbons and carbon monoxide remaining in the exhaust.

Alphanumeric A numbering system consisting of a combination of letters and numbers. They are placed in order starting from the left digit and working across to the right. This system is commonly used in parts catalogs and price sheets.

Alternating current (AC) An electrical current which flows alternately in two directions, forward and backward. It is produced by some form of mechanical device or motion, such as an alternator. Alternating current cannot be used to charge a battery. It must first be converted (rectified) to direct current for battery charging.

Alternator The electrical device driven by the engine which recharges the battery. It produces alternating current using rotating field coils inside stationary stator windings.

Anti-lock brakes A brake system which operates by pulsing the pressure to the wheel and caliper cylinders to prevent the lockup of any one wheel, thus preventing a skid.

Automatic transmission A transmission which automatically selects the correct gear ratio needed for the driving conditions. The driver has only to select the direction of travel desired.

Automotive aftermarket The distribution and sale of automotive replacement products.

Back order Merchandise ordered from a supplier but not shipped due to the supplier being out of stock.

Ball joint Part of the steering system which connects the spindle to the control arm. It allows the steering knuckle to turn right and left as well as permitting the control arm to move up and down.

Band A flexible flat piece inside an automatic transmission which holds parts of the gearset to create the correct gear ratio.

Battery A device within the electrical system which stores voltage until it is needed for vehicle operation.

Bearing A term used for ball bearing; an antifriction device having an inner and outer race with one or more rows of hardened steel balls between them.

Bill of lading A shipping document acknowledging receipt of goods and stating terms of delivery.

Blend door Part of the air conditioning/ventilation system which controls the ratio of incoming heated air and fresh air. This ratio is adjusted to control the temperature of the passenger compartment.

Blowby The unburned fuel and combustion byproducts which leak past the piston rings and into the crankcase.

Brake pads The friction elements of a caliper-type brake system. They are usually forced toward the rotor by hydraulic pressure.

Brake shoes The friction elements of a drum-type brake system. They are usually expanded outward by hydraulic pressure to contact the inside diameter of the brake drum.

Brake spring pliers Pliers used in the disassembly of brake drum components.

Break point Where cost of shipping by a particular method changes significantly because of size or weight classifications. For example, parcel post shipments cannot exceed 40 pounds.

Bushing A smooth cylinder used to reduce friction and to guide the motion of the parts.

Camshaft The shaft in an engine which causes the valves to open and close at the correct times.

Capacitor An electrical device used to smooth out changes in voltage levels.

Carbon dioxide A colorless, odorless, incombustible gas which is exhaled by humans and required by plant life.

Carbon monoxide A colorless, odorless, toxic gas formed by internal combustion engines.

Carburetor A device used on some engines to mix the fuel and air in the correct ratio for efficient combustion.

Cash discount A discount given for the prompt payment of a bill.

Cataloging The process of looking up the needed parts in the parts catalog.

Catalytic converter A metal canister mounted in the exhaust system containing metals that convert harmful exhaust gases into safer gases. A catalytic converter speeds up the reaction but is not consumed in the chemical reaction.

Caustic A compound that is able to burn, corrode, or eat away another compound.

Celsius The metric unit of temperature measurement.

Check valve A valve which allows something to move or flow in only one direction.

Circuit breaker An electrical device which protects the circuit by interrupting the current flow when it exceeds the rated capacity. It may be reset manually or automatically.

Closed loop An operating state in a computer-controlled engine in which the computer is controlling engine operation based upon information created by the sensors.

Closing a sale Getting a commitment to buy from a customer.

Clutch (1) The part of a driveline system which is used to interrupt the power flow between the engine and the transmission. (2) A part mounted on the air conditioning (AC) compressor used for temperature control of the AC system.

Clutch-release lever The part of a manual transmission system which operates the clutch-release (throwout) bearing.

Coil spring A spring which is wound into a spiral shape. Coil springs are commonly used on automotive suspension systems.

Combustion chamber The area inside the cylinder head and block where the burning of the fuel takes place.

Compression The act of forcing something together. In an automotive engine, the air and fuel are compressed in the cylinders to create a larger explosion and thus more power upon ignition.

Compressor The part of the air conditioning system which compresses the refrigerant vapor and pumps the refrigerant.

Computer terminal A keyboard system that permits a parts specialist, counter person, or operator to input invoices or obtain information from a computer usually located in a warehouse or distribution offices.

Condenser The part of the air conditioning system which cools the hot vapor and converts it to a liquid. The condenser is usually mounted in front of or on top of a vehicle for better airflow.

Constant-Velocity (CV) joint Part of the drivetrain which allows for changes in the angle of a driveshaft or half shaft.

Control arm Suspension parts which control spring action and the direction of travel of the axle as it reacts to driving conditions.

Coolant A fluid used for cooling, usually consisting of a blend of water and antifreeze.

Core Items such as starters, alternators, carburetors, and brake shoes accepted in exchange for remanufactured items.

Core charge A charge which is added when the customer buys a remanufactured part. Core charges are refunded to the customer when the core is returned.

Correction bulletin A bulletin which corrects catalog errors due to printing errors or inaccurately assigned part numbers.

Counter cards Advertising or display placards, usually with an easel back, placed on the counter.

Countershaft A shaft used in transmissions to transfer the motion from the input shaft to the output shaft.

Crankcase The lower part of an engine.

Crankshaft The shaft in an engine which converts the reciprocating piston motion into rotary motion for the driven device.

Credit memo The record of an amount paid to a customer, usually for the return of a purchased item or a core.

Crossover pipe The pipe which connects the two exhaust pipes of a dual exhaust system.

Cubic inch The volume equal to a cube with one-inch sides. The term is commonly used to describe engine displacement.

Customer relations A description of how a salesperson interacts with the customer.

Cylinder head The removable part covering the top of the engine cylinders. It seals the combustion chamber and usually contains the valves and spark plugs.

Dating Payments for merchandise extended to 30, 60, or 90 days without loss of cash discounts.

Dealers The jobber's wholesale customers, such as service stations, garages, and car dealers, who install parts in their consumers' vehicles.

Demand Items Items such as water pumps, bearings, clutches, and remanufactured parts that a customer needs for a specific vehicle.

Deposit A specified sum of money a customer leaves with the jobber for special orders or to guarantee the return of a core.

Desoccant Any substance used to absorb moisture. Desiccant is used in an air conditioning system to keep moisture from forming corrosive compounds.

Diameter The distance straight across a circular figure; the largest measurement which can be taken across a circular object.

Differential The set of gears which transmits power from the driveshaft to the wheels and allows the drive wheels to turn at different speeds for cornering.

Direct current (DC) An electrical current which flows in only one direction. It is usually created from a chemical source, such as a battery, and can be used to recharge a battery.

Direct mail Advertising literature sent to customers through the mail.

Discount A deduction from the stated price, usually a percentage off the printed price sheet.

Display A special grouping of sale items or new product lines with point-of-sale materials designed to attract customers' attention.

Distributor (1) The part of the ignition system which directs the secondary voltage (spark) to the correct spark plug. (2) A large volume stocking parts house that sells to wholesalers.

Do-it-yourselfer Someone who performs service and repair on his or her own vehicle.

Drag link The rod which connects the steering box to the steering knuckle on a straight axle front.

Driveshaft The shaft which connects the transmission to the differential.

Dual account A classification of a customer according to the purposes of the purchases. For example, a dual account customer can be one who both uses and sells parts, or a customer who uses parts for industrial and automotive purposes. The term is used in wholesaler policies that affect discounts, taxes, or credit.

Dual exhaust An exhaust system which uses separate exhaust pipes for each bank of cylinders.

Early Fuel Evaporization (EFE) An emissions system used to warm the air going into a cold engine to improve driveability and reduce hydrocarbon and carbon monoxide emissions.

Emergency order An order placed out of the normal cycle of stock orders, usually when unexpected or emergency parts are needed.

End cap A display fixture located at the end of gondolas.

Engine block The main body of an engine. The block contains the cylinders and carries the accessories.

Engine coolant The solution used in an engine to carry heat away from the cylinders. It is usually a 50/50 mixture of ethylene glycol and water.

Glossary

Evaporator The part of an air conditioning system which is mounted inside the passenger compartment. It contains low-pressure refrigerant liquid and vapor to remove heat from the air forced through and around it.

Excise tax Taxes on products paid by manufacturers or sellers usually included in the price of an item, but often added, as in the case of tires.

Exhaust Gas Recirculation (EGR) An emissions device which meters some exhaust gas into the intake manifold to reduce the combustion chamber temperature and to prevent the formation of oxides of nitrogen.

Exhaust manifold The part of the exhaust system which bolts directly to the cylinder heads. Exhaust gases leaving the cylinder are routed through the cylinder head to the exhaust manifold.

Extra dating A discount is made available for purchased items that are delivered at a given date and marked payable in a given period, such as 30, 60, or 90 days. It is used to spread payment without loss of the cash discount.

Facing Rotating a product so its labels are facing toward the front of the shelf. Facing improves shelf appearance and customer appeal.

Fahrenheit The temperature unit used in the English system of measurement.

Fastback A vehicle body style which has a fixed rear glass. Access to the luggage compartment is from the inside of the vehicle.

Fixtures The furnishings in a store such as gondolas, endcaps, display cases, racks, shelving, and counters.

Flare A type of tubing connection which requires that the end of the tubing be expanded at a 37 or 45 degree angle.

Flasher The part of the lighting electrical system which causes the turn signals and hazard lights to blink.

Fleet A number of vehicles operated by one owner or company.

Franchise A specified sum of money an individual must pay for the privilege of owning and operating one of a chain of retail operations.

Freight bill A bill that accompanies goods shipped describing the contents, weight, point of origin, shipper, and giving the transportation charges.

Freight charge A charge added to special order parts to cover transportation to the store.

Fuel injector The fuel system component which sprays fuel into the intake manifold on vehicles not equipped with a carburetor.

Fuel pump The fuel system component which moves the fuel from the tank to the carburetor or the fuel injection unit.

Fuse An electrical device which protects the circuit by interrupting the current flow when the flow exceeds the fuse's rated capacity. Fuses are constructed of a conductor encased in plastic or glass and must be replaced when blown.

Fusible link An electrical device which protects the circuit by interrupting the current flow when the flow exceeds the rated capacity. A fusible link is constructed of a wire surrounded by a special insulation and must be replaced when blown.

General application items Products such as oil, polish, antifreeze, and chemicals that apply to all makes of vehicles.

Gondolas Long, shelved display fixtures usually placed back to back to divide stores into aisled trafficways.

Grade (1) A method of classifying bolt strength. The grade of a bolt is indicated by markings on the head. (2) The classification of a product, such as motor oil.

Gross profit The selling price of an item less the cost.

Half shaft One of two shafts which connect a transaxle to the drive wheels.

Handling charge The cost charged to a customer for returning an item to a supplier or manufacturer for repairs or adjustment.

Hatchback A vehicle body style in which the rear glass will lift to allow access to the trunk.

Headers A tubular form of exhaust manifold which is used to increase exhaust gas flow and improve performance.

High-volume Describes a popular item which is sold in large quantities.

Hydrocarbon An organic compound made up of hydrogen and carbon. Most automotive fuels and lubricants contain hydrocarbons, a common source of pollution. As an automotive term, it refers to unburned fuel in the exhaust.

Hydrogen A colorless, highly flammable gas which is the most common in the universe. It is used in the production of methanol, an automotive fuel and fuel additive.

Idle A condition in which the engine is running at a low speed.

Idler arm A part of the steering system which connects the center link to the frame.

Ignition coil A part of the ignition system which produces the electricity needed to create the spark for the spark plugs.

Ignition system The part of the automotive electrical system which creates and delivers the high-voltage spark to the spark plugs.

Impulse item Any item that the customer did not intend to purchase when entering the store.

Inboard A position reference relating to being toward the center, or inside, of the vehicle.

Individually priced The condition of having the price of each part on the display marked for the customer's convenience.

Input shaft The shaft which carries torque into the transmission.

Insert bearings A split circular shell which is inserted between the engine block or connecting rods and the crankshaft.

Intake manifold The part of the intake system which bolts directly to the cylinder heads. The air or air/fuel mixture must pass through the intake manifold before entering the cylinder head and combustion chamber.

Intake valve The valve in the intake port which opens to allow the air/fuel mixture to enter the combustion chamber.

Intercooler A part of the intake system used with a turbocharger to cool the air entering the intake manifold.

Integrated circuit A miniaturized electrical component-consisting of diodes, transistors, resistors, and capacitors-used in electronic circuits.

Interchange list A cross reference for part numbers of identical items from different manufacturers.

Intermediate order An order placed after a regular stock order to replenish stock until the regular stock order shipment is received.

Inventory The goods a store has in its possession for resale.

Inventory control A method of determining amounts of merchandise to order based on supplies on hand and past sales.

Invoice A sales slip used as a record of sales.

Invoice register A system used to maintain a check on the issuing of invoices.

Jobber The owner or operator of an auto parts store usually wholesaling products to volume purchasers such as dealers, fleet owners, and industrial firms, and retailing to do-it-yourselfers.

Journal The part of a shaft which is supported by and comes in contact with a bearing.

Julian A calendar system devised by Julius Caesar which numbers the days consecutively starting with January 1. The Julian calendar is often used in production codes and computer systems.

Lifter A component of the automotive valve train which converts the rotary motion of a camshaft lobe into the reciprocating motion needed to open and close the valves.

Limited-slip differential A differential which uses internal clutch plates to limit the slip and speed difference between the drive wheels. This limiting improves traction on slick surfaces and helps eliminate wheel spinning.

Line The product list of a specific manufacturer.

List price The suggested selling price to the final consumer.

Liter A unit of volume used in the metric system of measurement. A liter is slightly greater than a quart in the English system.

Locator pad A quick-reference chart which will list a part's description, on-hand quantity, cost, and location in part number order.

Lost sale A customer not purchasing a requested item either because the store does not stock the item or because the store stocks brands other than the one requested by the customer.

Machine work Work done in a machine shop. In automotive applications, such work refers to engine block boring, shaft journal turning, brake drum and rotor turning, and cylinder head repair services.

MacPherson strut A suspension system component which combines a lower lateral link with a vertical strut to combine the features of a spindle and a shock absorber.

Main bearing caps The caps which secure the crankshaft to the block.

Maintenance free A term generally associated with a battery that requires no water or charging under normal conditions for the length of the warranty.

Margin Same as gross profit: the cost subtracted from the selling price.

Marketing The total function of moving merchandise from the manufacturer into the hands of the final consumer; including buying, selling, transporting, storing, and advertising.

Mark-up The amount a merchant charges for merchandise above costs to earn a profit.

Material Safety Data Sheet (MSDS) A prepared printed sheet which accompanies chemicals and contains information regarding the proper application along with the safety and environmental concerns.

Media A vehicle used for advertising, such as newspapers, magazines, television, radio, and outdoor posters.

Merchandising The "second effort" of advertising, such as building store displays of advertised items, posting newspaper advertisements and sales brochures, and promoting events through public relations.

Micrometer A precision measuring tool which can measure to dimensions of 0.0001 inch (English) or similar dimensions in the metric system.

Microprocessor A small computer-like device used to process signals from various circuits to achieve a control system that will adapt to operating changes.

Modules Self-contained electronic circuits for computerized systems; usually repaired by plugging in a new unit.

Muffler The part of the exhaust system which is used to reduce exhaust noise.

Net price The cost of an item to a particular purchaser, such as dealer net, jobber net, or user net.

Net profit The merchant's profit after deducting costs of merchandise and all expenses involved in operating the business.

No-return policy A store policy that certain parts cannot be returned after purchase. It is very common on electrical and electronic parts.

Obsolescence When a part is no longer of use, either because of replacement by a superseding part, or due to lack of demand.

OE An abbreviation for original equipment.

OEM An abbreviation for original equipment manufacturer.

Oil filter A device on the engine for removing dirt, carbon, and other impurities from the lubricating oil.

On-hand The quantity of an item that the store has in stock.

Operating capital The money that a merchant needs in everyday operations to buy parts, pay salaries, and meet regular expenses.

Open loop An operating state in a computer-controlled engine in which the computer is controlling engine operation based upon a predetermined program. It is usually in effect until the engine sensors signal that the engine has reached operating temperature.

Orifice tube A part of an air conditioning system which causes the drop in refrigerant pressure that results in the change in temperature necessary for cooling.

OSHA An abbreviation for Occupational Safety and Health Administration. All companies with a specified minimum number of employees must comply with its safety regulations.

Outboard A position reference indicating a state of being toward the outside of the vehicle.

Overage More items received than ordered.

Overhead The costs of operating a business not including purchases of merchandise.

Overhead cam An engine valve train system which has the camshaft positioned on top of the cylinder head.

Oxides of nitrogen (NO_x) By product of combustion responsible for photo-chemical pollution. Develops in high temperature combustion chamber. The primary purpose of EGR systems is to reduce Oxides of nitrogen.

Oxygen A gas making up approximately 18 percent of the air in the atmosphere and required for the combustion process to take place in an engine.

Oxygen sensor A computer system sensor which monitors the oxygen content in the exhaust gas. The signal from the oxygen sensor tells

Glossary

the computer whether a rich or lean condition exists in the air/fuel ratio entering the engine.

Packing slip A list of items and quantities accompanying a shipment.

Paid out A form or slip used to record such cash register transactions as returns or refunds.

Parking brake A mechanical brake on the vehicle used for parking or emergency stopping situations.

Parts specialist A person who is trained and skilled in selling parts at a parts counter.

Percentage of profit Profit accruing on the basis of the selling price of the product.

Perpetual inventory A method of keeping a continuous record of stock on hand through sales receipts or invoices.

Physical inventory Determining stock on hand through an actual count of items.

Pilferage Theft; shoplifting.

Pilot bearing A bearing mounted in the center of the crankshaft which supports the end of the transmission input shaft.

Pinion yoke The yoke mounted on the end of the pinion gear of the differential. The pinion yoke transfers torque from the driveshaft to the pinion gear.

Point-of-purchase (POP) materials Advertising materials used to promote sales.

Point-of-sale (POS) materials Advertising materials such as window banners, placards, and counter cards used at a place of business.

Policy A guiding rule for the conduct of a business.

Port fuel injection An automotive fuel delivery system which has a fuel injector for each cylinder positioned in the intake manifold at the base of the intake valve.

Positive crankcase ventilation (PCV) An emissions system which draws hydrocarbons from the engine crankcase and routes them through the intake manifold to be burned in the engine.

Price leader An item advertised at a very low price to attract customers into a store.

Price sheets Price lists usually accompanying catalogs from manufacturers or warehouse distributors; often in different colors for different types of customers.

Professional One who performs a specialized service as a means of employment, such as an ASE-certified automotive technician.

Profit The amount received for goods or services above the amount of expenses.

Public relations Providing the media with stories of interest worthy of being reported in the news.

Purchase order A form from a buyer that specifies items desired, conditions of sale, and terms of delivery.

Pushrod A valve train component used in engines to connect the lifter to the rocker arm.

R-12 The trade name for a refrigerant commonly used in automotive air conditioning systems. It has been phased out because of its hazard to the earth's ozone layer. R-12 is being replaced by R-134a as an automotive refrigerant.

R-134a The trade name for a refrigerant currently used in automotive air conditioning systems. It is the replacement of choice for R-12.

Rack and pinion A steering system which uses a horizontal rack with gear teeth and a pinion gear attached to the end of a steering shaft.

Radiator The device mounted at the front of the vehicle which is used to cool the engine. The hot coolant flows through the radiator. The radiator fins contain the coolant, and air flowing past the fins will remove heat.

Rain check A coupon that guarantees a customer the advertised price of a sold out item when it is again available.

Rebuilt A remanufactured part or assembly.

Reciprocating ball A steering system which uses a worm gear filled with ball bearings to reduce steering effort.

Rectifier An electrical device which converts alternating current (AC) into direct current (DC). The AC current produced by the alternator must be converted to DC in order to charge the battery.

Refrigerant A substance used to carry heat away from an object. The term usually refers to the chemical used in air conditioning and refrigeration systems.

Related items The group of items and tools-such as oil, filter, drain pan, filter wrench, and pour spout-needed to perform a particular job on a car.

Relay An electrical device which allows the remote control of a switch. It normally allows the control of a large current with a much smaller current.

Remanufactured part A part which has been reconditioned to original standards.

Reserve stock Merchandise usually kept in the storeroom to restock shelves and displays in the display area.

Resonator A type of secondary muffler.

Restocking fee The fee charged by the store or supplier for having to handle a returned part.

Retail Selling merchandise to walk-in trade, the do-it-yourselfers.

Return policy The policy established by the store regarding the return of unwanted or unneeded parts. Return policies may include restocking fees or prohibit the return of certain types of parts, such as electrical or electronic components.

Returns Merchandise returned by customers, usually for a refund or exchange.

Right-hand rule Store layout designed to permit customers to move naturally to the right on entering.

Rod bearing caps Rod bearing caps hold the rod to the crankshaft.

Rotor A part of the brake system which turns with the wheel spindle and is clamped by the brake pads for braking action. It is a term that also refers to the rotating part of an alternator that contains the field windings.

SAE An abbreviation for Society of Automotive Engineers, which sets the standards for many products.

Salvage That part or parts of an item retrieved from total loss or scrap which is still suitable for restoration, use, or resale.

Seasonal item An item which appeals to the customer during only a certain part of the year. Ice scrapers and lawn equipment are examples of seasonal items.

Selling price The price at which merchandise is sold, but might differ according to the type of customer (wholesale or retail).

Selling up Selling a customer a better quality item when a lower priced item obviously will not do the job he or she expects of it.

Service bulletin A bulletin which provides supplemental and update information for service manuals.

Service manual A manual which contains the diagnosis and repair procedures for vehicle systems.

Servo The hydraulic piston which applies the bands found in an automatic transmission.

Shelf talkers Small placards or flags placed with the promoted items on display shelves to call attention to sale prices, seasonal items, or new products.

Shoplift The act of stealing goods or display items from a store.

Solvent A chemical which is able to dissolve another substance. Solvents are normally used in the automotive industry as cleaning agents.

Spark plug The device which ignites the air/fuel mixture in the combustion chamber.

Special order An order placed whenever a customer purchases an item not normally kept in stock.

Spot A radio or television commercial.

Stand-up A tall, self-supporting point-of-sale piece, usually flat and printed.

Staple item Items such as oil, coolant, additives, and car care products stocked in the display area that customers regularly purchase.

Statement The bill, usually monthly, that a merchant receives for purchases, and also the bill a merchant sends to credit customers.

Stock order A process by which the store orders more stock from the suppliers.

Stock rotation Selling the older stock on hand before selling the newer stock.

Stud A fastening device which resembles a bolt. It has threads at each end and no head.

Sulfuric acid The acid used as an electrolyte in automotive batteries. It is corrosive and produces explosive hydrogen gas while being charged.

Supersession bulletin A bulletin sent by the parts supplier which lists part numbers that now supersede (replace) previous part numbers.

Supplements Catalog and price sheet changes used to keep items and prices current until new catalogs are issued.

Tabloid brochure A colorful, printed advertising piece used in retail promotions as a newspaper insert, direct mail piece, and for door-to-door delivery.

Technical bulletin A bulletin which explains unusual installation, fit, or maintenance problems associated with a part.

Thermostat (1) The device which controls the water flow through the radiator. It prevents water from flowing through the radiator until the engine is at operating temperature. (2) The device that controls the temperature of an air conditioning system.

Throttle body A fuel injection system which places the fuel injectors in a housing at the location previously used by the carburetor.

Throwout bearing A part of the drivetrain used with manual transmissions. It presses in the pressure plate fingers to release the clutch. It is also known as the clutch-release bearing and is mounted on the clutch release lever.

Thrust angle The alignment of the rear axle to the vehicle.

Tie-rod end The movable joint which connects the tie rods to the steering knuckle. Wear in the tie-rod ends will cause excessive tire wear.

Timing belt A belt which connects the camshaft and crankshaft, synchronizing their rotation. Timing belts are normally used with overhead cam engines.

Timing chain A chain which connects the camshaft and the crankshaft, synchronizing their rotation.

Timing gears The gears mounted on the end of the camshaft and crankshaft. They mesh together and synchronize the rotation of the camshaft and crankshaft.

Torque converter The device located between the engine and the automatic transmission. Hydraulically couples the engine crankshaft and automatic transmission input shaft and allows variable application of engine torque to transmission.

Torque multiplication The process of increasing the torque output of a torque converter. Torque converters usually provide some amount of torque multiplication for vehicle acceleration.

Trade discount A discount from the regular selling price offered to high-volume purchasers such as dealers, fleet owners, and industrial companies.

Traffic Customers.

Traffic builders All types of advertising, promotions, and merchandising materials designed to bring customers into a store.

Transaxle A combination transmission and differential.

Transfer case A gearbox which connects the front and rear drivelines of a four-wheel drive vehicle.

Transmission A set of gears which can change ratios to meet the various driving needs of the vehicle.

Transverse An orientation which refers to the engine and transaxle being mounted crosswise in the vehicle.

Trim tag A tag located on the vehicle body which outlines the color, body style, trim, and body accessories found on the vehicle.

Turbocharger A device mounted on the engine which forces air into the intake manifold. Turbochargers are driven by exhaust gas from the engine and serve to boost power and performance.

Turnover The number of times each year that a merchant buys, sells, and replaces a part number.

Universal joint (U-joint) A flexible coupling which connects the driveshaft to the pinion yoke of the differential.

Vehicle Identification Number (VIN) A unique number assigned to a vehicle for identification purposes. The VIN number can be decoded for information regarding year of manufacture, manufacturer, body style, engine size, carrying capacity, and other information.

Vendor The supplier.

Voltage regulator The charging system device which sets the maximum charging voltage produced by the alternator.

Walk-in Retail and do-it-yourself customers.

Glossary

Warehouse distributor (WD) The jobber's supplier. The link between manufacturer and jobber.

Warranty A printed document stating that a product will provide satisfactory service for a given period of time or the buyer will be entitled to a settlement according to the terms of the warranty.

Warranty return A defective part returned to the supplier due to failure during its warranty period.

Water pump The pump located at the front of the engine which circulates engine coolant throughout the cooling system.

WATTS Long-distance telephone service used by companies making a large number of long-distance calls daily.

Wheel bearings The bearings located between the wheel hubs and spindle or axle housing and the axle.

Wholesale The merchant's price to large volume customers such as dealers, fleet owners, and industrial companies. Unlike a trade discount, it is usually a fixed price rather than a percentage off the retail price.

Will-call Merchandise held for a customer to pick up.

Window banners Large posters displayed in windows to attract passersby to items on sale or promotions in progress.

Wire hangers Point-of-sale advertising materials attached to or draped over a wire inside the store.

Worm gear A gear cut in such a way that it spirals around a shaft.

Notes

Notes

Notes